ISBN: 9798558799514
Imprint: Independently published

2021

Project Engineering and Good Manufacturing Practices (GMP)

Des O'Brien

Contents

Units and Measurement
Introduction 6
SI Units and their Benefits 6
Linear Measurement 7
Manual Measuring Instruments 8
Slip Gauges 9
Vision Systems 10
Coordinate Measuring Machines 10
Optical Comparators 11
Geometric Tolerancing 11

Basic Statistics
Introduction 13
Inspection by Attributes 14
Test for Normality 14
Hypothesis Testing 14
Acceptance Sampling 15

Good Manufacturing Practices
Fundamentals 18
Quality 23
Basic Rules for GMP 24
Quality Management 33

Data Integrity
Introduction 43
System Categorisation 46

Facilities
Introduction 54
Clean Rooms 55
HVACS 57
Environmental Monitoring 62
Gown-Up Areas 65
Area Classification 66

Utilities
Introduction 70
Compressed Air 70
Water Systems 70

Sterile Manufacturing
Introduction 74
Container Closure 80
Depyrogenation 87
In-Process Controls 93

Validation
Equipment Qualification 97
The Validation Lifecycle 97
Process Validation 97
Dominance 100
Process Performance Qualification 101

Cleaning Validation
Introduction 102
Sampling 109

Units and Measurement

Introduction

This chapter covers the international system of units (SI units) and the basic principles of linear measurement. Communicating numbers and values is a basic requirement to allow engineering to occur. SI units provide a system that ensures different companies or sub-contractors across the globe can interrupt drawings and specifications.

SI Units and their Benefits

The Seven SI base units, which are comprised of:

m - Meter - Length
s - Second - Time
mol - Mole - Amount of Substance
A - Ampere - Electric Current
K - Kelvin - Temperature
cd - Candela - Luminous Intensity
kg - Kilogram – Mass

The benefit of applying the International System of Units (SI) is that written technical information is effectively communicated, overcoming the variations of language, including spelling and pronunciation. Arabic numerals describe the quantity. A quantity is then paired with a unit symbol, often with a prefix symbol that modifies unit magnitude.

Basic application of units and symbols

- Units: Names of units are made plural only when the numerical value that precedes them is more than one. Examples: 0.25 liter (quantity is less than one) and 250 milliliters (quantity is more than one).

- Symbols also remain the same despite the value. For example, 250 mm = 250 millimeters, is not 250 mms.

- The names of all units should start with a lower case letter except, at the beginning of a sentence. There is one exception: in "degree Celsius" (symbol °C) the unit "degree" is lower case but "Celsius" is capitalized. E.g. 37 degrees Celsius.

- Unit symbols are written in lower case letters except for liter and those units derived from the name of a person (m for meter, but W for watt, Pa for pascal, etc.).

- Symbols of prefixes that mean a million or more are capitalized and those less than a million are lower case (M for mega (millions), m for milli (thousandths).

- The dot or period (full stop) is used as the decimal point within numbers. In numbers less than one, zero should be written before the decimal point. Examples: 7.038 g; 0.038 g.

The prefix "kilo" stands for one thousand of the named unit. It is not a stand-alone term in the metric system. The most common misuse of this is the use of "kilo" for a "kilogram" of something. The word "micron" is obsolete and no longer used with "micrometer" the correct term. Also "degree centigrade" is no longer the correct unit term for temperature in the metric system; it has been replaced by degree Celsius.

The SI unit of time (actually time interval) is the second (s) and should be used in all technical calculations. When time relates to calendar cycles, the minute (min), hour (h), and day (d) might be necessary. For example, the kilometer per hour (km/h) is the usual unit for expressing vehicular speeds.

Linear Measurement

This involves the comparison of the piece under test with a known standard. These days, many manufacturing processes utilise automated measurement or complex vision systems. However, an understanding of the basic principles of measurement along with traditional measurement methods is fundamental to the engineer. Many of these traditional measurement methods are contact in nature (opposed to non-contact). While modern vision systems tend to be non-contact in nature, which offers multiple benefits.

Working standards are calibrated against master standards meaning that a comparison between the absolute standard and the particular measuring instrument is achieved. The calibration and testing of all measuring tools and instruments usually requires special equipment and should be completed in accordance with a recognised calibration standard.

Non-contact methods of measurement do not require physical contact with the part under measurement. The measurement is achieved by the use of optics, lasers or magnification. This can be an advantage as the part does not get exposed to physical forces during measurement. Non-contact methods also eliminates the "feel factor" and resulting human error of hand gages.

Contact measuring methods include micrometers and plug gages where accuracy and consistency are subject to the amount of pressure applied to the part during measurement.

Care & Maintenance of Instruments

1. Keep all instruments clean, treat them with care and avoid misuse.
2. Place instruments in cases or fit covers when not in use.
3. Keep the inside of the instrument case clean. The case is meant to protect the instrument.
4. Do not attempt to dismantle an instrument. If it is not functioning correctly return it to the appropriate department for servicing.
5. Choose the instrument in keeping with the tolerance of the dimension to be measured.
6. Wherever possible use an instrument that gives a direct reading.
7. Do not use worn or damaged instruments.
8. Remember that the graduations on a measuring instrument (resolution) are not necessarily the accuracy to which it can be used. Quite often the in-built inaccuracies of measuring instruments exceed their resolution.

Manual Measuring Instruments

The ease with which a component can be measured and the accuracy to which it need be or can be measured depends on the correct choice of measuring instrument. Factors such as the shape of the component and the position of the dimension to be measured influence the choice of instrument but nevertheless, certain basic rules should be followed.

Instrument or Measurement	Type of Measurement	Range	Value of Smallest Graduation (Resolution)	Suggested Reliability/ Accuracy
Steel rule	direct	150-500mm.	.5 mm.	±.5 mm.
Depth gauge	direct	150 mm.	.5 mm.	±.5 mm.
Callipers	direct	150 mm.	none	±.5 mm.
Vernier	direct	600	.01 mm.	±.05 mm.

callipers			mm.		
Vernier depth gauge	direct		300 mm.	.01 mm.	±.05 mm.
Vernier height gauge	direct		600 mm.	.01 mm.	±.05 mm.
Micrometres					
25-50 mm.	direct			.01 mm.	±.01 mm.
150-300 mm. plain	direct			.01 mm.	±.01 mm.
150-300 mm. plain	direct			.01 mm.	±.02 mm.
Inside micrometres	direct			.01 mm.	±.01 mm.
Depth micrometre	direct			.01 mm.	±.01 mm.
Telescopic gauges	transfer		150 mm.	none	±.02 mm.
Slip gauges	end standard		Up to 100 mm.	.001 mm.	±.0005 mm.
Dial test indicator	comparison		5 mm	.01 mm.	±.01 mm.
Dial test indicator	comparison		1.0 mm	.001 mm.	±.001 mm.

The volume of measurements is also a limiting factor. While a properly trained individual may be able to make repeated measurements, automation should be considered for repetitive inspection if possible.

Slip Gauges

Slip gauges or block gauges are used as standards for precision length measurement throughout the engineering industry. The gauges are usually made in sets and consist of a number of hardened steel blocks. Each block has two of the opposite faces lapped flat and parallel to a definite size within an extremely tight tolerance. In building up packs of slip gauges, errors can occur if care is not exercised. There are three main causes of errors:

1. Errors due to deviation from true size.
2. Errors due to grease or dirt between the wringing faces.
3. Errors due to expansion caused by excessive handling or leaving gauges exposed to strong sunlight or electric lamps.

Another manual instrument, the sine bar, which is one of the most effective methods of precision measurement of angles consists of a rectangular section bar to which are attached two hardened rollers of the same diameter, such that the common centre line of the rollers is parallel to the top face of the bar. The principle of the sine bar is based on the fact that in a right angled triangle the function known as the sine of the angles is the relationship of the hypotenuse to the side opposite the angle.

$$\text{Sin } A = \frac{a}{c}$$

To set up a sine bar to a required angle:
(a) Select and clean the appropriate bar.
(b) Determine the sine of the angle required by reference to the sine tables.
(c) Calculate the slip gauges required, i.e. multiply the sine of the angle by the size of the sine bar, i.e. the distance between the centres of the rollers.
(d) Select and clean the necessary slip gauges including protector slips.
(e) Wring the slip gauges together.
(f) Clean the surface plate and set up the bar by inserting the slip gauges under the roller as shown.

Sine centres are a very useful adaption of the sine bar principle. Adjustable centres are mounted on the face of the sine bar allowing such items as taper gauges and taper shafts to be inserted between centres and be measured for both angle and concentricity.

Vision Systems

While manual measurement via the previous instrumentation is useful in certain circumstances of low volume manufacturing, the requirements of high volume manufacturing and reliable quality calls for automated vision measuring systems to meet the volumes of inspection required.

Coordinate Measuring Machines

Coordinate measuring machines are available in a wide range of sizes and accuracy and can meet most precision 3D measuring applications for today's needs. Depending on the system setup and capability, a wide range of contact and non-contact probes can allow numerous kinds of measurements to be performed. CMM software assists in the analysis and interpretation of measurement results, which is particularly useful with increasing quantities of measurement.

Optical Comparators

Often referred to simply as a comparator, they are used to measure parts dimensionally by using optics and projection. Measurement is achieved by overlaying limits or graduations over the image projected. Inspection with comparators is relatively quick and is most useful when looking for a pass/fail result.

However, the rise of vision-based inspection systems makes the manual comparator, even equipped with modern capabilities, an oft overlooked tool. This is particularly the case when the requirement is to inspect large quantities of parts at once, since a vision system allows you to place multiple parts for inspection on the stage at the same time. For many simple measurements on two-dimensional parts with clearly defined edges, the optical comparator is quite suited.

Optical comparators can provide more information than just simple dimensions. Length and width measurements of the part shown above, for example, can be quickly obtained from two separate measurements by using a micrometer. These superficial measurements, however, might not reveal burrs, scratches, indentations or undesirable chamfers. Such imperfections are best detected on a comparator. In addition, a comparator's screen can be simultaneously viewed by more than one person and provide a medium for discussion, just as a white board might facilitate a conference.

Advantages
- Fast length and width measurement
- Length and width measurement can be done simultaneously
- Burrs and chamfers can be detected
- Screen can be viewed by more than one person

Digital comparators allow Pass/Fail inspection comparisons, Files are simply uploaded to the equipment in order to compare measurement results against a CAD overlay - the overlay moves with the datum, so you don't even need to use an optical comparator / profile projector.

Geometric Tolerancing

Before an object measured, complete information about both the size and dimensions and tolerances of the object must be available. The shape of an object is communicated through orthographic drawings, which are developed following standard drawing practices. The process of adding size information to a drawing is known as dimensioning.

Geometric Dimensioning and Tolerancing (GD&T) is a way of defining and communicating engineering tolerances. The geometry, tolerances and other information such as surface finish, concentricity or parallelism are expressed using symbols and text. In turn, properly created engineering drawings communicate the degree of accuracy and precision needed on each controlled feature of the part.

Tolerancing specifications define the allowable variation for the form and possibly the size of individual features, and the allowable variation in orientation and location between features. Two examples are linear dimensions and feature control frames using a datum reference (both shown above).

Standards define GD&T rules and drawing conventions include:

- American Society of Mechanical Engineers (ASME) Y14.5-2009
- International Organization for Standardization (ISO) ISO 2768:1989 General tolerances
  -Part 1: Tolerances for linear and angular dimensions without individual tolerance indications
  -Part 2: Geometrical tolerances for features without individual tolerance indications
- International Organization for Standardization (ISO) ISO 286:2010 Geometrical Product Specification

However, it should be noted that ASME Y14.5 standard provides a fairly complete set of standards for geometric dimensioning and tolerancing in one document. Where the ISO standards address a single topic at a time under different standards.

Dimension: a numerical value that defines the size or geometric characteristic of a feature.
Basic dimension: a numerical value defining the theoretically exact size of a feature. ' Reference dimension is the numerical value enclosed in parentheses provided for information only and is not used in the fabrication of the part.
Leader line: a thin solid line used to indicate the feature with which a dimension, note, or symbol is associated.
Datum: a theoretically exact point used as a reference for tabular dimensioning.
Tolerance: the amount a particular dimension is allowed to vary.
Dimension line: a thin solid line which shows the extent and direction of a dimension. Arrows are placed at the ends of dimension lines to show the limits of the dimension.
Extension line: a thin solid line perpendicular to a dimension line indicating which feature is associated with the dimension.

Basic Statistics

Introduction

Data is used to inform engineers and make decisions on failure modes, corrective actions or as justifications to make adjustments to process parameters. The type of data (variable or attribute) mostly depends on how the characteristic or feature is measured and recorded. For example, a product dimension or characteristic may be a continuous variable feature (e.g. thickness of 2.200mm), but if a go/no-go gauge is used to disposition the characteristic it is recorded as attribute data.

Variable data is typically measured and recorded on a continuous scale that provides precision greater than the number of digits required by the technical specification.

Examples of variable data include:

- Dimension in mm of an orthopaedic fixing plate
- Tensile force of a spring in Newtons
- Total drug content of a substance

Attribute characteristics tend to be a pass/fail, presence or absence of a characteristic. Examples include:

- Optical lens is scratched (cosmetic defect)
- Part is laser marked or not
- The product defective or not
- The tablet is broken or not

Inspection by Attributes

This is a method of inspection wherein either the unit of product is classified simply as conforming or nonconforming, or the number of nonconformities in the unit of product is counted with respect to a given requirement or set of requirements. Defining a defect is fundamental to sampling by inspection for attributes. There are two types of qualitative data- nominal and ordinal. Nominal represents categories that cannot be put in any order, while ordinal represents categories that can be ordered. There are also two types of quantitative levels - interval and ratio. They both represent "numbers" however, ratios have a true zero but intervals don't.

Test for Normality

Statistical methods such as t-test, ANOVA, regression analysis and as confidence and reliability are based on sampling from a population with specific parameters including the mean and the standard deviation. They are known as parametric methods. Parametric methods should conform to certain conditions, such as the requirement of normality for them to be suitable for use. Therefore, it is necessary to determine if the data distribution is normal or symmetric.

One such means of testing for normality is the Chi-Square test. This assesses a given data set and determines if the assumption of normality is true or false. Standardized Skewness and kurtosis also provides a test of normality to determine if the distribution is normal to use statistical tolerance limit calculations (confidence and reliability). Values of skewness and kurtosis outside the range of -2 to +2 indicate significant non normal data which would deem statistical tests regarding standard deviation unsuitable.

Hypothesis Testing

For statistical analysis two hypothesis must be considered. The hypotheses are defined as:

Null Hypothesis, Ho
Alternative Hypothesis, Ha

If the null hypothesis is rejected, the alternative hypothesis is concluded to be true. However, if the null hypothesis is not rejected, no conclusion is reached; the null hypothesis is not concluded to be true. The only way that a conclusion is reached is if the null hypothesis is rejected.

Acceptance Sampling

Acceptance sampling involves the inspection of product or raw material before in enters or as it leaves the factory. Its purpose is to ensure that defective products do not reach the patient or consumer. Similar to the type of data, there are two main types of sampling plans (1) attribute and (2) variable. Sampling plans involve either an AQL or RQL.

An acceptable Quality Level (AQL) is the defect rate you are willing to accept for a high percentage of the time.
The probability that a lot will be incorrectly rejected at the AQL. RQL is the probability that a lot will be incorrectly passed. An attribute acceptance sampling plan consists of a sample size and accept number. If the number of defects found exceeds the accept number, the lot is rejected. Variables acceptance sampling plans typically have smaller sample sizes than attribute acceptance sampling plans.

Statistical sampling and inspection is used for new product or service design, verification of those designs, incoming inspection, in-process inspection and final inspection. The level of sampling and inspection depends on the characteristics of the product, capability of the process, and the level of assurance required.

Lot inspection plans are determined based on the quality characteristic defect level and the manufacturing process used when assigning sampling plans and Acceptable Quality Level (AQL)

Pp (Process Performance) and Cp (Process Capability): quantify the stability of a process, i.e., the amount of variation in the output
Ppk (Process Performance Index) and Cpk (Process Capability Index) : These represent both the degree of variability AND the degree that the output is centred between the lower and upper specification limits.

Pp and Ppk are calculated using the standard deviation of the entire population and represent a long term performance. (Typically measured over a number of batches/runs and reflects both normal and special causes of variation). Pp and Ppk values are used to describe the Process expected over the long term.

Cp and Cpk are calculated using a sample standard deviation and represent a potential that could be achieved if sources of variation are eliminated.

Cp and Cpk are more commonly used during process optimisation studies as these represent the potential capability that could be achieved if the process were made stable by reducing special causes of variation. They may be applied during Operational Qualification. For stable processes (i.e. those where special cause variation is eliminated or minimised) Pp/Ppk and Cp/Cpk values will be similar in value.

Short term capability: Results from collecting data over a short period of time, (typically using only a single operator, a single manufacturing line, a single lot of material, etc.). **Typically expressed as Cp and Cpk.**

Long term capability: Based on reasonably independent observations collected from a process exhibiting normal production variation, for example using multiple operators, lines and lots of material over time. **Typically expressed as Pp and Ppk.**

Confidence Level

- Confidence Level is expressed as a percentage and represents the probability that the conclusion of the test is correct. A 95% confidence level means you can be 95% certain that the conclusion is correct.

- OQ typically should have a confidence level of 90%

- PQ typically should have a confidence level of 95%

AQLS

The AQL of a sampling plan is the Process Performance Level routinely **accepted** by the sampling plan.

- AQL is generally defined as the Process Performance Level that the sampling plan will **accept** 95% of the time - a 95% probability that the process will be accepted, i.e. the validation passes.

- This means validations for processes/equipment producing devices with a Process Performance Level at or better than the AQL are accepted at least 95% of the time and rejected at most 5% of the time.

    It describes the risk (5%) associated with rejecting a good process.

- The **producer** would like to design a sampling plan such that there is a **high probability** of **accepting** a validation

- The RQL of a sampling plan is the Process Performance Level routinely **rejected** by the sampling plan. RQL0.10 is defined as the Process Performance Level that the sampling plan will **reject** 90% of the time -**a 90% probability that the process will be rejected ,i.e. the validation fails.**

- This means that validations for processes/equipment producing devices with a **Process   Performance Level at or worse than the RQL are rejected at least 90% of the time and accepted   at most 10% of the time.**

- RQL0.05 is defined as the Process Performance Level that the sampling plan will **reject**   95% of the time.

- **It describes the risk associated with releasing/accepting a bad process.**

- The **consumer** would like the sampling plan to have a **high probability** of **rejecting** a validation with a Process Performance Level (% nonconforming) greater than, or equal to, the RQL.

Good Manufacturing Practices

Fundamentals

Good Manufacturing Practices are a set of practices that are required in order to comply with industry standards and regulations. GMP helps to minimise the risks involved during manufacturing and helps to ensure products meet quality and regulatory standards. A GMP quality system ensures that products are consistently produced and controlled according to predefined quality standards. It is designed to minimise the risks involved in any pharmaceutical production that cannot be eliminated through testing the final product.

Often, a broader term is used in industry -GxP-where the "x" is used as an umbrella letter representing different subjects or disciplines in industry. Some prime examples include GLP (Good Laboratory Practice), GDP (Good Documentation Practice), GEP (Good Engineering Practice) and GMP (Good Manufacturing Practices). Furthermore, the use of a lower case "c" as a prefix indicates "current" or "up-to-date". cGMP stands for "Current Good Manufacturing Practices". This means that some conventions or practices are subject to change within the industry. Therefore, it is important to be up-to-date in the application of cGxP or Cgmp. There are multiple regulators and organisations that provide definitions of "Good Manufacturing Practices". They include Organisations such as the World Health Organisation (WHO) and the International Society of Pharmaceutical Engineering (ISPE). Other definitions are offered by bodies such as the American competent authority for Food and Drug Administration. It is good to have an awareness of how organisations, bodies and competent authorities define GMP, and one should always review the "local" regulatory landscape. Below some definitions are provided to provide a feel for GMP and highlight the common thread between definitions.

W.H.O. World Health Organisation-"Good Manufacturing Practices (GMP, also referred to as 'cGMP' or 'current Good Manufacturing Practice') is the aspect of quality assurance that ensures that medicinal products are consistently produced and controlled to the quality standards appropriate to their intended use and as required by the product specification."

Food and Drug Administration: cGMP refers to the Current Good Manufacturing Practice regulations enforced by the US Food and Drug Administration (FDA). cGMPs ensure systems are properly designed and monitored, safeguarding the control of manufacturing processes and facilities. Adherence to the cGMP regulations ensures the identity, strength, quality, and purity of drug products by requiring that manufacturers of medications adequately control manufacturing operations. This includes establishing strong Quality Management Systems, obtaining appropriate quality raw materials, establishing robust operating procedures, detecting and investigating product quality deviations and maintaining reliable testing laboratories. This formal system of controls at a pharmaceutical company, if adequately put into practice, helps to prevent instances of contamination, mix-ups, deviations, failures and errors. This assures that drug products meet their quality standards.

MHRA (Medicines and Healthcare Products Regulatory Agency) defines GMP as follows:

"Good Manufacturing Practice (GMP) is that part of quality assurance which ensures that medicinal products are consistently produced and controlled to the quality standards appropriate to their intended use and as required by the marketing authorisation (MA) or product specification. GMP is concerned with both production and quality control. Many of the drivers of GMP in effect are also benefits to the manufacturer. Good manufacturing practices are an expected practice in regulated industries and a manufacturer must meet all relevant GMP regulations if they wish to manufacture within a country or sell to a particular market. It is important to maintain accurate, complete, up-to-date and consistent information to ensure patient safety and reduce any potential risks."

Documentation Creation

The principles of GDP should be applied at the document creation stage. As most people are familiar with softcopy or electronic documents, some of these points are obvious but nonetheless need to be made. All documents should be electronically written and not handwritten except for execution of protocols, test results and adding entries. Documents that are approved controlled should be:
Accurate and free from errors
Have revision or version controlled
Should have an effective date or date of release

Approval of Documents

Document approval must be completed by trained and appropriately experienced personnel. Often companies will use an approval matrix which explains which people are required to approve each document. For example, an EHS (Environment Health and Safety) officer would be required to approve a risk assessment.

Signatures

A signature on any document is legally binding so remember to read and understand what is being signed for. Every signature should also include the date in the correct format. If a signature appears within the same document alongside initials, substituting a full signature with initials and date is generally acceptable. This practice is common when large documents are being completed.

Date and Time Format

A standardised approach to dates and times is important especially within large global organisations. For instance, in the USA, the norm is to place the month before the date, whereas in Ireland and Great Britain it is common to write the day of the month followed by the month. Most companies would define their date and time format in an SOP or procedure.

The date and time format can also be configured in Word documents and Excel worksheets to align with a companies preferred date and time format.

Handwritten Entries

When a handwritten entry is required such as a signature or a test result, indelible ink must be used. Many companies will have an SOP or procedure that states the specific ink colour required. If an entry of a test result or test data isn't completed at the time of execution, this constitutes a late entry. Backdating an entry or signature is forbidden. Always use the correct and current date.

How Are Mistakes Corrected?

This is a critical area of GDP. Failure to follow the requirements of GDP when correcting mistakes is the most common failure when it comes to documentation in industry. The method of correcting mistakes using GDP allows for a person looking at the document for the first time to clearly see the original entry and the corrected entry. This maintains the integrity of the document. In order to identify the changes and corrections, certain rules must be followed. No overwriting is allowed and white-out or Tipp-Ex is not allowed.

Accuracy
Accuracy of information provided in documents is critical in the life science industry. As the end user is a patient, inaccurate records or documents could cause serious injury or death. Controlled documents are also legal documents and could be called upon if recalls, litigation or investigations arise.

Many documents used in the manufacture of medical devices are designed to record information or test results. These test results are then used to disposition (pass or fail) batches of product. Inaccurate information could risk the release and distribution of defective product. This has a potential impact on both the business and the patient or user.

Blank Spaces or Blank Fields

On completion of a document such as a logbook or record, no blanks spaces should be left unfilled. This is to avoid late entries and also to prevent confusion. Blank spaces or blank fields should have a diagonal line drawn neatly across the space, the letters "N/A" written and the entry signed and dated. If the reason for "N/A" is not evident then it is wise to include an explanatory note or sentence.

Data Transcription

Transcribing is the process of transferring data from one source to another. This is often required when raw data is involved. When data is in raw format it may need to be entered into a Microsoft Excel sheet. When transcribing data remember that all original raw data must be stored in case it is needed at a future date. After the data is transcribed it must be verified by a second person to check for any errors or omissions.

Revision Control

Controlled documents should always have a version number or revision number electronically on each page of the document. This is similar to books which always list what edition they are e.g. first edition or second edition. Revision control is a key element of the Quality Management Systems in place in regulated industries. As the need for changes in the document arises, the controlled document can be amended/updated. With each update the version number revises also. Some companies will use alphabetic revision control and to a lesser extent numeric revision control (Version A, Version B or Version 01, Version 02).

Management of Attachments

Attachments to controlled documents can include training records, data sheets, lab results and so on. It is important that attachments are identified for traceability purposes. If the attached becomes detached from the main document, the attachment should be identifiable. It is best practice to include a reference number on the attachment if available. If the attachment consists of several pages, each page should be numbered in Page X of Y format if not electronically done so. And remember, hand written entries must be accompanied by a signature and date. Always use staples to attach documents together. Glue or paper clips are not acceptable.

Management of Documents through Their Lifecycle

GDP applies to all the different stages of a document's lifecycle. These stages include creation, review, approval, issuing, completion of records, revision, updating, retirement and storage. Storage a.k.a. retention is an important stage and often a legal requirement for medical devices and pharmaceutical products. For consumer OTC medicines a 5-year retention of quality records often suffices. For implants such as TKRs or Total Knee Replacements, a 90-year retention period is required. This ensures that traceability and a quality record is available if the need arises.

Test Results

This section provides an overview on the correct handling of test results. Test results can be generated from various types of product testing such as visual inspection, dimensional inspection and chemical analysis. The recording of all test results should be completed on an approved form. This is to ensure that the correct information is being recorded and the same approach is taken by different people who might have to complete testing.

Calculations

There are different ways calculations can be completed. Many simple calculations can be done by an individual using a calculator, alternatively, a software package such as Minitab or an Excel sheet can be used to complete complex calculations. It should be clear to the reader what calculation is required, what the formula is and how the calculation is completed. If the formula used is not included on the sheet, it should be referenced in a controlled document. Care is also required when recording numbers of several decimal places in length, as rounding error can be introduced.

Quality

Quality can be defined as the ability to consistently produce products meeting the same specifications time after time. Products must be safe, pure, uniform and effective. Specifications can be set down internally within a company, however, depending on the product, external specifications from regulators or standards may be required.

Patient safety is the primary focus of any pharmaceutical drug or medical device. This is the expectation of any patient or user. Secondly, the patient or user is interested in receiving an effective product. It is product specifications that ensure these criteria are accounted for.

The key elements of a QMS are listed below. The ISO Standard, ISO 9001, is a global Quality Management standard used by thousands of organisations and companies. This standard sets out the requirements of a QMS.

Quality Policy: A company will document their commitment and approach to quality within their organisation. It usually sets out how they plan to achieve a high and consistent standard of quality. It should in some way speak to the customer or end user.

Quality Objectives: Quality objectives can be documented in a Quality Plan at site or organisational level. An effective way of defining quality objectives is use of the SMART method. SMART stands for Specific, Measurable, Achievable, Realistic and Timely.

Quality Manual: An in-house guidance document to provide a framework for achieving the quality objectives.

Organisational Structure and Responsibilities: Organisational charts can be used to map out the company structure. Roles and responsibilities can be documented in site quality plans, job descriptions and Standard Operating Procedures.

Data Management: A coherent approach to the provision, storage and maintenance of data.

Processes: Processes are defined and documented.

Resources: Resources must be properly understood, allocated and linked across the organisation.

Product Quality & Customer Satisfaction: The proper management and investigation of complaints is important to reduce future instances from reoccurring. Continual engagement with the end user or customer is critical.

Continuous improvement including corrective and preventive action- where continuous improvement projects and initiatives are encouraged and supported. The application of a CAPA system to ensure quality is maintained and consistent.

Maintenance: A Preventative Maintenance schedule is in place and managed accordingly.
Sustainability: All work practices are sustainable and consistent throughout the lifecycle of processes and products.
Auditing: Systems are auditable and maintained to allow internal or external review and audit.
Engineering Change Control: Where changes are required to validated processes or equipment, changes are managed and introduced under change control.

A common acronym used to highlight the aims of Good Manufacturing Practices (GMP) is SPUE which stands for Safe-Pure-Uniform-Effective. This definition is particularly suited to pharmaceutical products as the chemicals and drugs used need to be pure and free of contaminants. Furthermore, they need to be uniform, meaning they will have the same constituents from tablet to tablet and batch to batch. A description of each word is shown below:

SAFE- the product has the right ingredients if it is a drug product. It is packaged as intended and correctly labelled in order to provide identification and safe use.

PURE- it is free of contaminants, foreign matter, chemicals and harmful microbes.

UNIFORM- The product is manufactured consistently and will have the same quality between batches manufactured on different days.

EFFECTIVE- Ultimately, the product must be effective in treating the medical condition. To be effective, it requires the correct ingredients, the correct amount of ingredients and correct packaging to maintain the product stability over time.

Basic Rules for GMP

Rule #1: Get the facility design right from the start

Rule #2: Validate processes

Rule #3 : Write good procedures and follow them

Rule #4 : Identify who does what

Rule #5: Keep good records

Rule #6 : Train and develop staff

Rule #7 : Practice good hygiene

Rule #8 : Maintain facilities and equipment

Rule #9: Build quality into the whole product lifecycle

Rule #10: Perform regular audits

Rule #1: Get the facility design right from the start

Every food, drug, and medical device manufacturer aims to operate their business in accordance with the principles of Good Manufacturing Practice (GMP). It's much easier to be GMP compliant if the design and construction of the facilities and equipment are right from the start. It's important to embody GMP principles and use GMP to drive every decision.

Facility layout

Lay out the production area to suit the sequence of operations. The aim is to reduce the chances of cross contamination and to avoid mix-ups and errors. For example, don't have final product passing through or near areas that contain intermediate products or raw materials. Aim to:

- remove unnecessary traffic in the production area
- segregate materials and products
- minimise potential for mix-ups and errors

Facilities Design

The following points should be considered at the facility design stage. The impact of choices and decisions here can must be understood. The scope and type of manufacturing will determine many of the building and facility requirements.

Risk assessments should be considered as a tool in identifying the right materials and design features.

Materials of construction
Windows
HVAC requirements
Utilities to be supplied
Emergency generators/UPS systems
Access to site and area's
Entrances

FDA Requirements

211.42 Design and Construction Features
(a) "Any building or buildings used in the manufacture, processing, packing, or holding of a drug product shall be of suitable size, construction, and location to facilitate cleaning, maintenance, and proper operation."

211.46 Ventilation, Air Filtration, Air Heating and Cooling

(b) "Equipment for adequate control over air pressure, micro-organisms, dust, humidity, and temperature shall be provided when appropriate for the manufacture, processing, packing, or holding of a drug product."

Equipment

Design, locate, and maintain equipment to suit its intended use. The equipment should be:

- easy to repair and maintain
- designed and installed in an area where it can be easily cleaned
- consistent with the intended use
- calibrated at defined intervals (if required)

Environment

It's important to control the air, water, lighting, ventilation, temperature, and humidity within a plant so that it does not impact product quality. You should design facilities to reduce the risk of contamination from the environment.

Make sure that:
- lighting, temperature, humidity and ventilation are appropriate
- walls, floors and ceilings are smooth, free from cracks and do not shed particulate matter
- interior surfaces are easy to clean
- pipe work, light fittings, and ventilation points are easy to clean
- drains are sized adequately and have trapped gullies.

Rule #2: Validate processes

Validation is defined as "Establishing documented evidence that provides a high degree of assurance that a specific process will consistently produce a product meeting its pre-determined specifications and quality attributes." It's a GMP requirement to prove control of the critical aspects of certain operations. New facilities and equipment, as well as significant changes to existing systems, require validation.

All validation activities should be well planned and clearly defined. This is usually by means of a Validation Master Plan, or VMP. Before you get to this stage consider all the critical parameters that may be affected and impact product quality; what happens if the stirring speed is changed? How does this affect temperature or pH? Once this is complete, define the testing and documentation required.

Validation usually is made up of three components:

- Installation Qualification, or IQ, which is testing to verify that the equipment is installed correctly as per manufacturers recommendations.

- Operational Qualification, or OQ, which is testing to verify that the equipment operates correctly as described in the user requirement specification.

- Performance Qualification, or PQ, which is testing to verify and confirm that product(s) can be consistently be produced to specification under anticipated conditions.

The FDA defines 4 types of validation. (1) Prospective, (2) Concurrent, (3) Retrospective and (4) Revalidation. These various types of validation form what approach the validation takes. E.g. is it a new process that will be validated in advance of commercial manufacturing, or, will the process be validated in a staged basis – concurrently. Etc. The 4 types of process validation are explained in the definitions section.

Rule #3 : Write good procedures and follow them

Within a regulated company, many documents are used to instruct, track, test and record information on the manufacturing process. Any document that can impact the quality of the product or product safety is treated as a controlled document. A controlled document is classified as a legal document. These controlled documents must incorporate certain requirements such as the date of approval, revision control and appropriate levels of review and approval. The accuracy and content of these documents can be subject to review by regulatory bodies Including the FDA in the US and the MHRA in the UK It is important that there are no errors or "questions marks" over the content. Examples of controlled documents include:

- Policies
- SOPs
- Specifications
- MFR (Master Formula Record)
- BMR
- Validation protocols and reports
- Forms
- Logbooks
- Records
- Bills of Materials (BOMs)

- Test Methods

Written procedures are controlled documents that provide detailed step-by-step instructions for the user. Written procedures promote consistency as they allow the same task to be performed in the same way, even by different people. They also act as a reference. If changes or improvements are identified, having a procedure in place creates a clear starting point which can be improved or modified in a controlled manner.

-Procedures should be written using clear and concise language
-Steps should be numbered clearly and individually to make them easy to follow
-Remember, written procedures are only effective if they are followed correctly, consistently and at all times by everyone

Never deviate from written procedures, these controlled documents ensure consistency and accuracy is maintained over time.

Rule #4: Identify roles

All employees understand what tasks and activities need to be done each day. Documenting these responsibilities is key to ensuring people fulfil their roles and responsibilities. Job descriptions for each role and employee should be created and should detail the following:

- job title
- job objective
- duties and responsibilities
- skill requirements

An organisational chart helps to document and display roles and functions.

Rule #5: Keep good records (Documentation)

Good science starts with good documentation. As part of GMP it is essential to keep accurate records, and during an audit, it helps convey that you are following procedures. It also demonstrates that processes are known and under control. Guidelines on Good Documentation and record keeping:

- Record all information immediately upon completion of a task
- Write legibly with indelible ink. By signing records you are certifying that the record is correct and that you have performed the task as per the defined procedure

- Correct mistakes as per GDP. Draw a single line through any error, and initial and date the correction. Include a reason for the correction at the bottom of the page if the reason is not obvious

- Record details if you deviate from a procedure. Ask your supervisor or the Quality Department for advice should a deviation occur

- If it's not documented then it didn't happen!

Rule #6: Train staff

Training should be provided for all employees who work within manufacturing, production or laboratories or where they may have an impact the quality of the product. Basic training on GMP is a fundamental requirement, followed by any job specific training as required. All training should be documented with a record of when the training occurred, dates, attendees and results of any assessments. Training records are often drawn upon by external auditors and are an essential record.

Personnel are central to the application of CGMP and compliance to regulations. At every level throughout an organisation, people interact with materials, equipment and processes in order to deliver products to the market and patient. Personnel must therefore be suitably qualified and equipped to carry out their responsibilities effectively.

Provisions in guidance and regulations are therefore made for personnel in a Quality Management System. Despite advances in automation and computerised systems, people are centrally involved in day-to-day decisions. For this reason, there must be sufficient and suitably qualified personnel to carry out all the tasks. Individual responsibilities should be clearly defined and understood by the persons concerned and recorded formally in procedures and job descriptions.

It may be an obvious point; however, manufacturers must ensure an adequate number of personnel with the necessary qualifications and practical experience are resourced to manufacturing. Having a broad base of people with the experience, knowledge and skills reduces the risk of quality issues. Responsibilities placed on any one individual should not be so extensive as to present any risk to quality.

Personnel should have specific duties recorded in written descriptions and adequate authority to carry out their responsibilities. Their duties may be delegated to designated deputies with a satisfactory level of qualifications.

Personal Hygiene

All personnel should be trained in the practices of personal hygiene. A high level of personal hygiene should be observed by all those concerned with manufacturing processes. Personnel should be instructed to wash their hands before entering production areas. Signage should be in place along with hand washing facilities. Hand washing demonstrations and training should be provided by a suitably qualified QC analyst or microbiologist. Any person experiencing an illness or exhibiting open lesions or wounds that may adversely affect the quality of products should not be allowed to handle starting materials, packaging materials, in-process materials or medicines until the condition is no longer a risk to quality or patient safety. Direct contact should be avoided between the operator's hands and starting materials, primary packaging materials and intermediate or bulk product.

Rule #7: Practice good hygiene

In order to reduce the risk of product contamination, good hygiene by every person is required. A culture of hygiene awareness should be evident in a GMP compliant facility. The practice of good hygiene should be supported by procedures and monitoring programs.

Depending on the classification of medical device been manufactured, the level of cleanliness required increases as the risk to patient increases. Therefore, sterile product manufacturing will require more strict controls and levels of hygiene.

Points to note:
- Practice good personal hygiene by washing your hands at regular intervals
- Wear the required PPE and protective garments and follow gowning procedures
- If you are ill, inform your manager
- Minimise any contact with product or product contact surfaces and manufacturing equipment.
- Do not eat or drink in GMP areas or where indicated

Rule #8: Maintain facilities and equipment

Preventative maintenance plays an important role in ensuring facilities and equipment remain fit for purpose. Regular maintenance prevents equipment breakdowns and unplanned interventions which disrupt production and cause backlogs within the process. Proper maintenance also reduces the risk of product contamination and helps to maintain the 'validated state' of the equipment and facility. GMP requires accurate records relating to maintenance activities are kept for audit and quality purposes.

Rule #9: Build quality into the whole product lifecycle

Quality must be paramount in any medical device. From its very design through development and into commercial manufacturing, quality must meet acceptable levels. Quality is the responsibility of everyone, not just the quality department.

Quality by Design

The traditional old school of thought focused on 100% inspection where defects are identified by operators. This approach to product quality leads to undetected defects and risk patient safety. This is not to say 100% inspection is not valuable, however, it is more effective is suitable automated systems conduct the inspection.

Online Control or in process control often uses statistical and process controls measures during the manufacturing stage to monitor and respond to drift in process settings and react if defects are produced or detected. The alternative to traditional inspection is to ensure quality is "built in" to the design of a product – Quality by design. Quality by design starts early on within the design and development stage. Potential defects or failure modes are identified and can be designed out of the product or controls put in place to reduce risks or communicate issues.

Control of Components

All materials and components when accepted onsite must meet predefined acceptance criteria. Sampling and accepting testing may be required.

Typically, suppliers provide certificates of conformance to ensure the materials or components meet specifications. Suitable storage conditions need also be accounted for. All materials and components must be approved prior to release for manufacturing.

Rejected materials or materials that fail inspection should be identified and stored in a secure area to prevent unauthorized use.

Control of the Manufacturing Process

Establish records and procedures to ensure that employees perform the same job every time. Each product must have:
- A master record that outlines the specifications and manufacturing procedures.
- Individual batch records to document conformance to the master record.
- SOPs and procedures for cleaning and maintaining the equipment and areas.

Packaging and Labelling Controls

Proper Packaging and labelling is necessary to identify how materials are packed, stored and labelled. Distinctive labels and accurate descriptions help prevent mix-ups and errors. Labelling also supports traceability, to different batches or different products.

Rule #10: Perform regular audits

Audits are conducted to assess if GMP rules and regulations are followed. External bodies such as the Food and Drug Administration (FDA) conduct formal audits also known as external audits. A Corrective Action Preventative Action (CAPA) system is required to manage and fix issues identified during an audit.

Simply put, an audit is a review activity that examines if a company or organizations processes are been followed. It also allows the identification and improvement of any concerns or non-compliances. Audits can be internal (conducted by internal staff) or external, (external-regulatory audits or certification bodies)

Audits are a key element of a quality management system. The process of establishing an internal audit process can be aided with reference to ISO 19001. This standard provides guidance and lots of examples on implementing and maintaining audit systems.

Benefits of Auditing

Audits provide a means of assessing a company's quality management system, and how well it is in compliance with the processes and procedures within the company.
Some key benefits of audits:
- Audits help verify compliance and conformity to requirements laid out in regulations and industry legislation. (e.g. ISO, FDA, Eudralex etc.)
- They measure the effectiveness of the QMS and the engagement of Top Management
- Audit help identify opportunities for improvements
- Audits promotes awareness of the Quality Management System

Quality Management

Systematic process for the assessment, control, communication and review of risks to the quality of the drug (medicinal) product across the product lifecycle. Achieving an effective product design, requires in depth knowledge of the customer requirements, clinical or medical need, regulatory requirements, and the manufacturing technology to be used.
This collective knowledge or "knowledge space" ensures a robust and quality product is more likely to be designed that will meet the market requirements. Literature, engineering studies and the qualification and experience of employee's all contribute to the knowledge space.
From this knowledge space, the most stable and effective design should be selected with product quality and safety as key factors. Furthermore, the quality of the product is then controlled and maintained within what can be described as the control space.
During the design stage of product development, specifications are created to describe the attributes and features of the product. The output of the design stage is to have the required product specification documents available as inputs to equipment selection and process selection.
Examples of some specifications include Raw material specifications, intermediate product specifications and finished product specifications. Specifications contain information on various features and product attributes such as dimensions, formulation, purity, cleanliness, surface finish and so on.

The critical requirements stated in specification are often referred to as Critical Quality Attributes (CQAs) and Critical Process Parameters (CPPs)
Critical Quality Attributes (CQA): a particular property of a material, product or output of a process that is key to the product performance and safety.
Critical Process Parameters (CPP): a process parameter such as temperature or time that when varied it impact the quality or CQA of a product.

Quality Control is that part of Good Manufacturing Practice which is concerned with sampling, specifications and testing, and with the organisation, documentation and release procedures which ensure that the necessary and relevant tests are actually carried out and that materials are not released for use, nor products released for sale or supply, until their quality has been judged to be satisfactory.

PICS/s Manufacturing Principles for Medicinal Products:

Pharmaceutical Inspection Convention and Pharmaceutical Inspection Co-Operation Scheme (PIC/S): The Pharmaceutical Inspection Convention and Pharmaceutical Inspection Co-Operation Scheme (jointly known as PIC/S) develop international standards between countries and pharmaceutical inspection authorities, to provide a harmonised and constructive co-operation in the field of Good Manufacturing Practices. PIC/S provides an active and constructive cooperation in the field of GMP and related areas. The purpose of PIC/S is to facilitate:

- ➢ networking between participating authorities
- ➢ maintenance of mutual confidence
- ➢ exchange of information and experience
- ➢ mutual training of GMP inspectors.

The guide consists of an introduction section along with two parts and a number of annexes.

- **Guide to Good Manufacturing Practice for Medicinal Products –** Introduction
    o Introduction
    o Adoption and entry into force
    o Revision history

- **Guide to Good Manufacturing Practice for Medicinal Products - Part I**

  Part I covers GMP principles for the manufacture of medicinal products

  1. Quality management
  2. Personnel
  3. Premises and equipment
  4. Documentation
  5. Production
  6. Quality control
  7. Contract manufacture and analysis
  8. Complaints and product recall
  9. Self-inspection

- **Guide to Good Manufacturing Practice for Medicinal Products - Part II**

  Part II covers GMP for active substances used as starting materials

  1. Introduction
  2. Quality management
  3. Personnel
  4. Buildings and facilities
  5. Process equipment
  6. Documentation and records
  7. Materials management
  8. Production and in-process controls
  9. Packaging and identification labelling of APIs and intermediates
  10. Storage and distribution
  11. Laboratory controls
  12. Validation
  13. Change control
  14. Rejection and re-use of materials
  15. Complaints and recalls
  16. Contract manufacturers (including laboratories)
  17. Agents, brokers, traders, distributors, repackers and relabellers
  18. Specific guidance for APIs manufactured by cell culture / fermentation
  19. APIs for use in clinical trials
  20. Glossary

The annexes provide detail on specific areas of activity and are listed below:

- **Technical interpretation of PIC/S GMP guide Annex 1** - Manufacture of sterile medicinal products

  PIC/S has published a recommendation for the technical interpretation of Annex 1 on the manufacture of sterile medicinal products.

  This recommendation summarises the interpretations an inspector adopts during an inspection of the manufacture of sterile medicinal products. It reflects the most important changes introduced in the revised Annex 1, but is not intended to address all changes in the revision.
  - Document history
  - Purpose and scope
  - Basics
  - Definitions and abbreviations
  - New texts and their interpretation
  - Revision history

- **Guide to Good Manufacturing Practice for Medicinal Products – Annexes**
  - Annex 1 - Manufacture of sterile medicinal products
  - Annex 2 - Manufacture of biological medicinal products for human use
  - Annex 3 - Manufacture of radiopharmaceuticals
  - Annex 4 - Manufacture of veterinary medicinal products other than immunologicals
  - Annex 5 - Manufacture of immunological veterinary medical products
  - Annex 6 - Manufacture of medicinal gases
  - Annex 7 - Manufacture of herbal medicinal products
  - Annex 8 - Sampling of starting and packaging materials
  - Annex 9 - Manufacture of liquids, creams and ointments
  - Annex 10 - Manufacture of pressurised metered dose aerosol preparations for inhalation
  - Annex 11 - Computerised systems
  - Annex 12 - Use of ionising radiation in the manufacture of medicinal products
  - Annex 13 - Manufacture of investigational medicinal products
  - Annex 14 - Manufacture of products derived from human blood or human plasma
  - Annex 15 - Qualification and validation
  - Annex 16 - Qualified person and batch release
  - Annex 17 - Parametric release
  - Annex 18 - GMP guide for active pharmaceutical ingredients (This annex no longer exists)
  - Annex 19 - Reference and retention samples

- o Annex 20 - Quality risk management
- o Glossary

EudraLex - Volume 4 - Good Manufacturing Practice (GMP) Guidelines

Volume 4 of the rules governing medicinal products in the European Union contains guidance for the interpretation of the principles and guidelines of good manufacturing practices for medicinal products for human and veterinary use laid down in Commission Directives 91/356/EEC, as amended by Directive 2003/94/EC, and 91/412/EEC respectively.

EudraLex V4 is made up of the following parts:

- ➢ Introduction
- ➢ Part I - Basic requirements for medicinal products
- ➢ Part II - Basic requirements for active substances used as starting materials
- ➢ Part III - GMP related documents

The Commission Directive 2003/94/EC, of 8 October 2003, sets out the principles and guidelines of good manufacturing practice in respect of medicinal products for human use and investigational medicinal products for human use.

Part I - Basic Requirements for Medicinal Products
- o Chapter 1 - Pharmaceutical Quality System
- o Chapter 2 - Personnel
- o Chapter 3 - Premise and Equipment
- o Chapter 4 - Documentation
- o Chapter 5 - Production
- o Chapter 6 - Quality Control
- o Chapter 7 - Outsourced Activities
- o Chapter 8 - Complaints and Product Recall
- o Chapter 9 - Self-Inspection

Part II - Basic Requirements for Active Substances Used as Starting Materials

Basic requirements for active substances used as starting materials.

Part III - GMP Related Documents
Site Master File
Q9 Quality Risk Management
Q10 Note for Guidance on Pharmaceutical Quality System
MRA Batch Certificate

Annexes
Annex 1- Manufacture of Sterile Medicinal Products
Annex 2- Manufacture of Biological Active Substances and Medicinal Products for Human
Annex 3- Manufacture of Radiopharmaceuticals
Annex 4- Manufacture of Veterinary Medicinal Products Other than Immunological Veterinary Medicinal Products
Annex 5- Manufacture of Immunological Veterinary Medicinal Products
Anne 6- Manufacture of Medicinal Gases
Annex 7- Manufacture of Herbal Medicinal Products
Annex 8- Sampling of Starting and Packaging Materials
Annex 9- Manufacture of Liquids, Creams and Ointments
Annex 10- Manufacture of Pressurised Metered Dose Aerosol Preparations for Inhalation
Annex 11- Computerised Systems
Annex 12- Use of Ionising Radiation in the Manufacture of Medicinal Products
Annex 13- Manufacture of Investigational Medicinal Products
Annex 14- Manufacture of Products Derived from Human Blood or Human Plasma
Annex 15-Qualification and Validation (in operation since 1 October 2015)
Annex 16- Certification by a Qualified Person and Batch Release
Annex 17- Parametric Release
Annex 19- Reference and Retention Samples

21 CFR Part 11 FDA

The FDA publishes regulations and guidance documents for industry in the Federal Register. The FDA's website also contains links to the cGMP regulations and guidance documents as well as various resources to help drug companies comply with the law. The FDA also conducts onsite audits and public outreach through presentations at national and international meetings and conferences on the subject of cGMP requirements.

Pharmaceutical quality affects every American. The FDA regulates the quality of pharmaceuticals very carefully. The main regulatory standard for ensuring pharmaceutical quality is the Current Good Manufacturing Practice (CGMPs) regulation for human pharmaceuticals. Consumers expect that each batch of medicines they take will meet quality standards so that they will be safe and effective. Most people, however, are not aware of CGMPs, or how the FDA ensures that drug manufacturing processes meet these basic objectives. Recently, the FDA has announced a number of regulatory actions taken against drug manufacturers based on the lack of CGMPs. This paper discusses some facts that may be helpful in understanding how CGMPs establish the foundation for drug product quality.

PART 211   Current Good Manufacturing Practice for Finished Pharmaceuticals

Subpart A--General Provisions
§ 211.1 - Scope
§ 211.3 - Definitions
Subpart B--Organisation and Personnel
§ 211.22 - Responsibilities of quality control unit
§ 211.25 - Personnel qualifications
§ 211.28 - Personnel responsibilities
§ 211.34 - Consultants
Subpart C--Buildings and Facilities
§ 211.42 - Design and construction features
§ 211.44 - Lighting
§ 211.46 - Ventilation, air filtration, air heating and cooling
§ 211.48 - Plumbing
§ 211.50 - Sewage and refuse
§ 211.52 - Washing and toilet facilities
§ 211.56 - Sanitation
§ 211.58 - Maintenance
Subpart D--Equipment
§ 211.63 - Equipment design, size, and location
§ 211.65 - Equipment construction
§ 211.67 - Equipment cleaning and maintenance
§ 211.68 - Automatic, mechanical, and electronic equipment
§ 211.72 - Filters
Subpart E--Control of Components and Drug Product Containers and Closures
§ 211.80 - General requirements
§ 211.82 - Receipt and storage of untested components, drug product containers, and closures
§ 211.84 - Testing and approval or rejection of components, drug product containers, and closures

§ 211.86 - Use of approved components, drug product containers, and closures
§ 211.87 - Retesting of approved components, drug product containers, and closures
§ 211.89 - Rejected components, drug product containers, and closures
§ 211.94 - Drug product containers and closures

Subpart F--Production and Process Controls
§ 211.100 - Written procedures; deviations
§ 211.101 - Charge-in of components
§ 211.103 - Calculation of yield
§ 211.105 - Equipment identification
§ 211.110 - Sampling and testing of in-process materials and drug products
§ 211.111 - Time limitations on production
§ 211.113 - Control of microbiological contamination
§ 211.115 - Reprocessing

Subpart G--Packaging and Labelling Control
§ 211.122 - Materials examination and usage criteria
§ 211.125 - Labelling issuance
§ 211.130 - Packaging and labelling operations
§ 211.132 - Tamper-evident packaging requirements for over-the-counter (OTC) human drug products
§ 211.134 - Drug product inspection
§ 211.137 - Expiration dating

Subpart H--Holding and Distribution
§ 211.142 - Warehousing procedures
§ 211.150 - Distribution procedures

Subpart I--Laboratory Controls
§ 211.160 - General requirements
§ 211.165 - Testing and release for distribution
§ 211.166 - Stability testing
§ 211.167 - Special testing requirements
§ 211.170 - Reserve samples
§ 211.173 - Laboratory animals
§ 211.176 - Penicillin contamination

Subpart J--Records and Reports
§ 211.180 - General requirements
§ 211.182 - Equipment cleaning and use log
§ 211.184 - Component, drug product container, closure, and labelling records
§ 211.186 - Master production and control records
§ 211.188 - Batch production and control records

§ 211.192 - Production record review
§ 211.194 - Laboratory records
§ 211.196 - Distribution records
§ 211.198 - Complaint files
Subpart K--Returned and Salvaged Drug Products
§ 211.204 - Returned drug products
§ 211.208 - Drug product salvaging

World Health Organisation GMP Guideline Annexes

The WHO Essential Medicines and Health Products (EMP) Department works with countries to promote affordable access to quality, safe and effective medicines, vaccines, diagnostics and other medical devices. As part of this effort, the WHO publishes a number of guidance annexes that describe best practice quality requirements for specific areas within the life science industry.

List of WHO GMP Annexes:
- WHO Good Manufacturing Practices for Pharmaceutical Products: Main Principles
  Annex 2, WHO Technical Report Series 986, 2014
- Active Pharmaceutical Ingredients (Bulk Drug Substances)
  Annex 2, WHO Technical Report Series 957, 2010
- Active Pharmaceutical Ingredients - Bulk Drug Substances: Additional Clarifications and Explanations
- Pharmaceutical Excipients
  Annex 5, WHO Technical Report Series 885, 1999
- WHO Good Manufacturing Practices for Sterile Pharmaceutical Products
  Annex 6, WHO Technical Report Series 961, 2011
- WHO Good Manufacturing Practices for Biological Products
  Annex 3, WHO Technical Report Series 996, 2016
- WHO Good Manufacturing Practices for Blood Establishments (jointly with the Expert Committee on Biological Standardisation)
  Annex 4, WHO Technical Report Series 961, 2011
- Pharmaceutical Products Containing Hazardous Substances
  Annex 3 WHO Technical Report Series 957, 2010
- Investigational Pharmaceutical Products for Clinical Trials in Humans
  Annex 7, WHO Technical Report Series 863, 1996
- Herbal Medicinal Products
  Annex 3, WHO Technical Report Series 937, 2006

- Radiopharmaceutical Products
  Annex 3, WHO Technical Report Series 908, 2003
- Water for Pharmaceutical Use
  Annex 2, WHO Technical Report Series 970, 2012
- WHO Guidelines on Good Manufacturing Practices for Heating, Ventilation and Air-Conditioning Systems for Non-Sterile Pharmaceutical Dosage Forms
  Annex 5, WHO Technical Report Series 961, 2011
- Validation
  Annex 4, WHO Technical Report Series 937, 2006
- Guidelines on Good Manufacturing Practices: Validation, Appendix 7: Non-Sterile Process Validation
  Annex 3, WHO Technical Report Series 992, 2015

International Council for Harmonisation, ICH, GMP Guide

The International Council for Harmonisation of (Technical Requirements) for Pharmaceuticals for Human Use (ICH) brings together the regulatory authorities and pharmaceutical industry to discuss scientific and technical aspects of drug registration. Since its inception in 1990, ICH has gradually evolved, to respond to the increasingly global face of drug development.

ICH Q7 Good Manufacturing Practice Guide for Active Pharmaceutical Ingredients

Data Integrity

Introduction

Data generated by or used in GxP impacting activities must be handled and protected in accordance with international and national regulatory requirements. The application of data integrity applies to many industries and products that touch the lives of patients and end users across the globe. Some examples of products that must meet data integrity regulations include (1) active pharmaceutical ingredients, (2) medical devices, (3) medicinal products, (4) vaccines and (5) cosmetics.

The below agencies and regulatory authorities provide specific requirements on data integrity:

- EU GMP – EudraLex – Rules Governing Medicinal Products in the European Union Volume 4 – Guidelines to Good Manufacturing Practice for Medicinal Products for Human Use – Products for Human and Veterinary Use, Annex 11: Computerised Systems – (1, 7.2, 17)

- FDA – 21 CFR Part 11 – Food and Drug Administration – Electronic Records; Electronic
Signatures – Scope and Application (C)

- FDA- 21 CFR Part 211 – Food and Drug Administration – Code of Federal Regulations -
Good Manufacturing Practices - 211.188a, 211.194.2, 211.194.8

- ICH E6 – International Conference on Harmonisation - Guideline for Good Clinical Practice
(5.2.1, 8.1, 8.3)

- MHRA – United Kingdom - Medicines and Healthcare Products Regulatory Agency - GMP
Data Integrity Definitions and Guidance for Industry (2015)

- PIC/S Guidance PI 011-] – Pharmaceutical Inspection Convention Scheme - Good
Practices for Computerised Systems in Regulated "GXP" Environments

Within the life science industry the saying goes "if it's not written down, it didn't happen". This is a powerful message that is a suitable starting point for data integrity. In the current and present day, the mere mention of data integrity quickly conjures an image of Excel sheets, big data, databases and computers in our minds. However, it has a broader impact with its roots in the basics of good science – good documentation.

Data integrity indeed does apply to "soft" or electronic data but also applies to paper-based systems and records. GxP is the umbrella acronym that stands of "good practices" in all our tasks and activities, be it laboratory testing, process engineering and so on. A core element in meeting GxP is abiding by "Good Documentation Practices" (GDP). Having good written records is fundamental to patient and product safety within the pharmaceutical, biopharmaceutical and medical devices industries. So, data integrity begins with the small stuff — real-time data collection, real-time review, honest and accurate recording of data and events.

The integrity of data relies on several factors. It can be influenced by a company's culture or approach to doing business. It can also be affected by the level of experience or knowledge within a company. Many traditional engineering companies outside the regulated life science community simply do not have the need to be so thorough in their handling of data and information. Within a GxP environment, controls, training and the design and operation of systems and processes influence data integrity on a day-to-day basis. Most of the time, those affected by the controls or systems do not think of them, but they can either support or inhibit data integrity and the reliability of data. Obviously, equipment, systems and processes should play a key role in making data reliable and accurate.

Regulations that speak to GxP and data integrity can apply to many different streams within the life science sector as previously mentioned. From medical devices to pharmaceuticals, all act in different manners, with long and short term applications. Take the example of a total knee replacement. Many designs now ensure their effectiveness in excess of ten years, even up to twenty years depending on individual circumstances. This requires many key records within manufacturing to be kept for several decades. Thus, data retention requirements specify the retention periods of such documents. The integrity of GxP data must be protected during the entire data life cycle, from creation of the data and records to the eventual destruction of data after the retention period is fulfilled. Data integrity equally applies to:

- Equipment
- Computerised systems

- Test records
- Inspection records
- Material certificates

Data integrity ensures that patient safety, product quality, and product supplies are generated by the product life cycle processes.

Process Design

Failure to maintain data integrity can occur throughout the life cycle of data; however, a thoughtful design of systems can prevent breaches in data and restrict the severity of any attempts to alter data. Therefore, design should aim to include controls and preventative measures. At a high level, this can be achieved by:

- Limiting access to GxP events and data
- Standard Operating Procedures (SOPs)
- Training
- System owners

Data Reliability

Data reliability is the foundation to achieving cGxP data integrity. The FDA's ALOCA model can be used to enforce data reliability.

Accuracy: the GxP data is recorded, calculated, analysed, and reported as found and correctly.

Attributable: any actions or calculations performed on GxP data can be attributed to or traceable to the person that performed the actions and the date and time at which they were performed.

Legible: the GxP data is recorded in a clear and human-readable form.

Contemporaneity: the GxP data is recorded at the same time as the observation/measurement is made or as soon as possible after the event.

Original: the initial data recorded is available and not altered.

An additional point to make it that of trustworthiness. It is assumed that engineers and scientists etc. working across the life science industries are ethical and do not falsify data or information. Typically companies can implement a code of practice or ethical behavior programmed to desist people from intentional unethical behaviour or the falsification of records.

System Categorisation

GAMP 5 makes provision for four categories of software in order to distinguish the level of customization/configurability that exists across software serving different functions

GAMP Software Category 1, Operating Systems

Category 1, operating systems, covers established commercially available operating systems. These systems are not subject to validation themselves. The name and version of the operating system must, however, be documented and verified during Installation Qualification (IQ). Application software hosted on operating systems needs to be validated.

GAMP Software Category 3, Non-Configured Software

Category 3 covers commercially available, standard software packages and "off the-shelf" solutions for certain processes. The configuration of the software packages should be limited to adaptation to the runtime environment (for example network and printer connections) and the configuration of the process parameters. The name and version of the standard software package should be documented and verified in an installation qualification (IQ). Special user requirements, such as security, alarms, messages, or algorithms must be documented and verified in an operational qualification (OQ).

GAMP Software Category 4, Configurable Software Packages

GAMP Software Category 4, Configurable Software Packages Category 4 covers configurable software packages that allow special business and manufacturing processes.
This involves configuring predefined software modules. These software packages should only be considered as belonging to Category 4 if they are well-known and mature. Normally, a supplier audit is necessary.
If this is not available, the software packages should be handled as Category 5. The name, version, and configuration should be documented and verified in an installation qualification (IQ).

The functions of the software packages should be verified in terms of the user requirements in an operational qualification (OQ). The validation plan should take into account the life cycle model and an assessment of suppliers and software packages.

GAMP Software Category 5, Custom Software

GAMP Software Category 5, Custom Software Custom/Bespoke Software (GAMP Software Cat 5) is software that contains custom code designed or modified specifically for a particular customer. As the code is custom, it presents a greater risk. This risk must be mitigated with the right approach to the validation.

Correctly assigning a GAMP software category to equipment, systems or processes is an important activity that should be completed early on in the planning stage of a project. There must of some degree of familiarity with the equipment or system. The manufacturer or vendor can be a source of information that may help the designation. In many cases, companies create tools or processes that help determine what GAMP software category applies. These have different names such as questionnaires, screening tools, planning tools etc.

21 CFR Part 11

This section specifically covers the regulatory requirements of part 11 of Title 21 of the Code of Federal Regulations; Electronic Records; Electronic Signatures (21 CFR Part 11). Part 11 of the FDA CFR is relevant to "records in electronic form that are created, modified, maintained, archived, retrieved, or transmitted under any records requirements set forth in agency regulations."
As of 2007, several sections of the regulation have been identified as excessive and the FDA announced in guidance that it will exercise enforcement discretion on some parts of 21 CFR part 11. This has been welcomed by some manufacturers but it has also caused a degree of confusion. The requirements relating to access controls are the most fundamental requirements and are routinely enforced. The "predicate rules" that required organisations to keep records in the first place are still in effect. If electronic records are illegible, inaccessible, or corrupted, manufacturers are still subject to those requirements.

If a regulated firm keeps "hard copies" of all required records, those paper documents can be considered the authoritative document for regulatory purposes. This then means that the computer system is not in scope for electronic records requirements, although subject to predicate rules which still require validation. If the "hard copy" is to be identified as the authoritative document, the "hard copy" must be a complete and accurate copy of the electronic source. The manufacturer must use the hard copy (rather than electronic versions stored in the system) of the records for regulated activities.

Definition of Records

The FDA has deemed the following records or signatures in electronic format subject to 21 CFR part 11:

Records that are required to be maintained under predicate rule requirements and that are maintained in electronic format in place of paper format. On the other hand, records (and any associated signatures) that are not required to be retained under predicate rules, but that are nonetheless maintained in electronic format, are not part 11 records. Records that are required to be maintained under predicate rules, that are maintained in electronic format in addition to paper format, and that are relied on to perform regulated activities.

Records submitted to FDA, under predicate rules (even if such records are not specifically identified in agency regulations) in electronic format (assuming the records have been identified in docket number 92S-0251 as the types of submissions the agency accepts in electronic format). However, a record that is not itself submitted, but is used containing nonbinding recommendations in generating a submission, is not a part 11 record unless it is otherwise required to be 205 maintained under a predicate rule and it is maintained in electronic format.

Electronic signatures that are intended to be the equivalent of handwritten signatures, initials, and other general signings required by predicate rules. Part 11 signatures include electronic signatures that are used, for example, to document the fact that certain events or actions occurred in accordance with the predicate rule (e.g. approved, reviewed, and verified).

The above definitions are taken from the FDA guidance document entitled "FDA Guidance for Industry: 21 CFR Part 11 - Electronic Records and Electronic Signatures: Scope and Application, August 2003." This document also provides recommendations on documenting key decisions that may be taken in relation to 21 CFR Part 11 applicability and compliance.

Requirements and Specifications

The need for compliance to 21 CFR depends on the type of technology and level of automation and computerisation involved in the manufacturing process or other actives that are GxP-impacting. Does the system store electronic records? Does the system require a login? Is there an audit trial? If a complex system is to be procured, the requirements need to be communicated to the manufacturer as part of a user requirement specification and/or software requirement specification.

General Guidance on Requirement Specifications

While the quality system regulation states that design input requirements must be documented, and that specified requirements must be verified, the regulation does not further clarify the distinction between the terms "requirement" and "specification." A requirement can be any need or expectation for a system or for its software. Requirements reflect the stated or implied needs of the customer, and may be market-based, contractual, or statutory, as well as an organisation's internal requirements.

There can be many different kinds of requirements (e.g., design, functional, implementation, interface, performance, or physical requirements). Software requirements are typically derived from the system requirements for those aspects of system functionality that have been allocated to software. Software requirements are typically stated in functional terms and are defined, refined, and updated as a development project progresses. Success in accurately and completely documenting software requirements is a crucial factor in successful validation of the resulting software. *Page 6 Guidance for Industry and FDA Staff General Principles of Software Validation A Specification* is defined as "a document that states requirements." (21 CFR 820.3(y)). It may refer to or include drawings, patterns, or other relevant documents and usually indicates the means and the criteria whereby conformity with the requirement can be checked.
There are many different kinds of written specifications, e.g., system requirements specification, software requirements specification, software design specification, software test specification, software integration specification, etc. All of these documents establish "specified requirements" and are design outputs for which various forms of verification are necessary.

Validation of Computerised Systems

The requirement for computerised systems to be compliant to 21 CFR part 11 needs to be identified early on in the project to ensure that the vendor or supplier of the systems or equipment can develop and build a system that meets the requirements of 21 CFR part 11. Computer system validation can be divided into three distinct phases: (1) planning, (2) design and development, (3) verification and (4) retirement.

Planning: This phase involves the planning of the validation effort required to implement the system and identification of key milestones and requirements. It requires supplier assessments, assessments of the regulatory and system risks, supplier development of a validation approach and the identification of deliverables that will be generated to support the implementation and operation of the system.

Design and Development: This phase consists of the design, development and configuration of the hardware and software required to meet the system requirements. In the case of custom software, design and developmental testing is important to ensure proper functionality prior to verification testing.

Verification: This phase confirms that requirements and specifications have been met. Testing is required to ensure the system operates as intended. Upon successful testing and verification, the system can be released for use. Once verification activities have begun, any changes to the system must be managed through change control. In case of successful completion of the verification activities (i.e. any deviation has been evaluated and addressed), the system is released for effective use. The operation phase supports the need to maintain compliance and fitness for intended use after the system is accepted and released for use.

Retirement: This phase consists of the planning, executing and summarising of the events required for system shutdown. It includes the appropriate handling of the supporting documents and the data contained within the system. While described here as a separate phase, a system's retirement can be handled as part of a new system implementation or as a separate project.

Best practice when it comes to computer system validation is to adopt a life cycle approach which requires the completion of activities in a systematic way from system conception to retirement. Life cycle activities could be scaled according to system impact on product quality, patient safety and data integrity, system complexity and novelty, supplier assessment and business risk.

Electronic Records

When it comes to the regulated industries such as the medical device industry, every process and procedure must be documented. Documentation ensures that everyone is working in the same manner with the same procedures. However, documentation is more than just writing down procedures and processes. It is also concerned with how documents are controlled, how they are updated and how they are stored.

Electronic Document management systems

Electronic document management systems aka EDMS are now the norm and gold standard for most medium to large organisations. Many companies that provide medical device manufacturers with an EDMS that can be customised to match the business processes particular to an organisation. With configurable or customisable software, validation and proper verification is important to ensure the system operates as intended. There are also regulatory requirements that stipulate the expectations and requirements of such systems. For example, the application of electronic signatures and the presence of audit trials. FDA 21 CFR Part 11 details the requirements with regard to electronic records and electronic signatures. For medicinal products in Europe, GMP V4 Annex 11 specifies similar requirements.

Record Retention

With regard to the part 11 requirements for the protection of records to enable their accurate and ready retrieval throughout the records retention period (11.10 (c)), persons must also comply with all applicable predicate rule requirements for record retention and availability such as (211.180(c) general requirements. The decision to follow 21 CFR part 11 should be justified and documented as part of a risk assessment and based on the value of the records over time. The FDA does not object to archiving of required records in electronic format to non-electronic media such as paper, or to a standard electronic file format (examples of such formats include, but are not limited to, PDF, XML, or SGML). Persons must still comply with all predicate rule requirements, and the records themselves and any copies of the required records should preserve their content and meaning. As long as predicate rule requirements are fully satisfied and the content and meaning of the records are preserved and archived, you can delete the electronic version of the records. In addition, paper and electronic record and signature components can coexist as long as predicate rule requirements are met and the content and meaning of those records are preserved.

Electronic Signatures

Electronic signatures are computer-generated character strings that count as the legal equivalent of a handwritten signature. The regulations for the use of electronic signatures are set out in 21 CFR Part 11 of the FDA. Each electronic signature must be assigned uniquely to one person and must not be used by any other person. It must be possible to confirm to the authorities that an electronic signature represents the legal equivalent of a handwritten signature. Electronic signatures can be biometrically based or the system can be set up without biometric features.

Conventional Electronic Signatures

If electronic signatures are used that are not based on biometrics, they must be created so that persons executing signatures must identify themselves using at least two identifying components. This also applies in all cases in which a chip card replaces one of the two identification components. These identifying components, can, for example consist of a user identifier and a password. The identification components must be assigned uniquely and must only be used by the actual owner of the signature.

When owners of signatures want to use their electronic signatures, they must identify themselves by means of at least two identification components. The exception to this rule is when the owner executes several electronic signatures during one uninterrupted session. In this case, persons executing signatures need to identify themselves with both identification components only when applying the first signature. For the second and subsequent signatures, one unique identification component (password) is then adequate identification.

Audit Trail

Title 21 CFR details predicate rule requirements relating to documentation of, for example, date time, or the sequencing of events, as well as any requirements for ensuring that changes to records do not obscure previous entries. Making the decision on whether to apply audit trails, or other appropriate measures, or on the need to comply with predicate rule requirements should involve a justified and documented risk assessment. Any risk assessment should determine the potential effect on product quality and safety and the integrity of the record.

Change Management

Validation programmes are subject to change control. Each company or organisation should have a procedure detailing the change management process.
Any system, facility, document or process that has the potential to impact product quality and the validated state is generally subject to following a change control process. Another term used in industry is enterprise change control or engineering change control. Essentially, these terms are the same. The intent is to control and manage change consistently.

A change control can take the form of a document which drives the agenda and the specific requirement. Change control is also created with enterprise software such as Kintana, Documentum and SAP. While each company will have varying processes, some basics are common. These include the three stages of change control; pre-implementation, implementation and post implementation (if required).

Validation Deliverables

The deliverables of validation activities should be in accordance with a project validation plan of validation master plan. For small projects or changes to computerised systems, a change control may serve as the validation plan. However, some typical deliverables include the following:

- GxP assessment (note, some systems may be non GxP applicable)
- User requirements specification
- Third party audit
- Validation plan
- Design specification such as functional, software, hardware and technical specifications
- GxP risk assessment
- Validation protocols
- Traceability matrix
- Validation report

Facilities

Introduction

Facilities and utilities qualifications are typically prerequisites to the validation of manufacturing equipment and systems. Much of the activity that deals with establishing a facility or building that is fit for purpose is managed under the broad heading of commissioning and qualification (C&Q). The terms C&Q are often used interchangeably and in practice some overlap in activity is expected. Commissioning can be defined as the planned, documented, and managed engineering approach to the start-up and handover of facilities, systems and equipment to the end-user. It must deliver a safe and functional environment that meets the predefined design and user requirements.

In strict terms, qualification is more concerned with the confirmation and documentation showing that equipment or systems are properly installed and functional. Qualification forms part of validation, but the individual qualification steps do not equal a validated process. The establishment of a user requirements specification (URS) and detailed design specifications ensure that the building or facility will meet end-users' needs and that it is fit for the intended purpose.

It also provides a level of protection to the contracting company responsible for the project or facility construction. Post-URS approval requires an approved Design Qualification (DQ). This provides verification and a documented record that the proposed design is suitable for the intended purpose. Further verification including IQ/OP/PQ should be applied as required based on the system impact and criticality of facilities/utilities.

A risk-based qualification process should assess the potential of a system to impact the product quality. The boundaries of any system (HVAC, compressed air supply etc.) should be identified in order to help establish the scope of any system and determine if it has a direct, indirect or no impact on product quality.

Direct Impact: a system that can directly impact product quality.

Indirect Impact: where a system is not expected to directly impact the product quality but supports or is ancillary to a direct impact system.

No Impact: a system that does not directly impact product quality and does not support a direct impact system.

Clean Rooms

All locations on earth except latitudes near the equator experience seasonal temperature changes. The changes are a consequence of Earth's orbital motion about the sun, coupled with the tilt of its axis of rotation with respect to its orbital plane. Design criteria should be based on published temperature data. The HVAC system design should consider the following:

Standard Operating Conditions: These are climatic conditions against which the systems must be designed to operate, control, and maintain required conditions. (These may be based on published data, which are only exceeded 2.5% or 1% of the time).

Extreme Operating Conditions: These are climatic conditions against which the systems must be designed to operate, without manual intervention, and without damage to the systems or the facility. Based on product / process risk assessments, extreme or standard conditions shall be used for HVAC design for dedicated areas.

Location
Based on the building layout, footprint and design intent, a suitable and adequate space must be identified for HVAC location. This must include provision of chilled water, heating systems, ducts and drainage. HVAC plants must be accommodated in designated HVAC plant rooms or interstitial areas.

Air Intake
During the design phase, the air intake locations should be selected to ensure air is in the best environmental condition. The below considerations help to achieve a strong starting point:

Thermal Load
Thermal load can be defined as the amount of heat energy to be removed from an inner environment by equipment (HVAC) used to maintain that environment at the design temperature when worst case external temperature(s) are being experienced. The thermal load requirement should be calculated for the following:
- Max summer conditions
- Minimum winter conditions
- High rainfall
- Standard operation
- Extreme operating conditions

Room Recovery Time
Room recovery time to return to the required pressure differential and cleanliness

Dust, Vapour, or Fume Control
Highlight areas requiring dust, vapour, gas and/or fume control on the room data sheet. These areas must be controlled to remove the possibility of product contamination and to ensure the safety of the operator and environment. Areas requiring 100% fresh air or extraction to atmosphere may require greater airflow or other measures within the room to maintain environmental conditions.
In order to meet the appropriate level of cleanliness, HVAC systems require sufficient filtration to provide "clean" air to prevent contamination of the product. Pre-filters and main filters are normally suitable for most operations; however, HEPA filters are required to prevent particulate or microbial contamination for higher-classification areas

Air Change Rates
The air change rates for each room must be calculated to be sufficient for clean-up to achieve specified particulate conditions "at rest" in static conditions after a maximum of 20 minutes from completion of operations. The actual air change rate must be chosen to satisfy the most stringent requirements including GMP, GLP, heat gain, ventilation requirements and/or occupancy, including an appropriate safety factor.
The air change rate must be optimised for energy savings; however, specific attention must be paid to air locks where a greater air change rate must be applied. Air changes can be reduced (e.g. setback modes) in some circumstances ("at rest" mode, with no production activity and no personnel interventions).

Room Environmental Conditions
Other environmental conditions to be controlled, such as temperature and relative humidity, depend on the product and nature of the operations carried out in those areas.
These parameters should not interfere with the defined cleanliness standard.

Temperature Requirement
The normal operating temperature requirement for each classification. Temperature and humidity must be appropriate to the product and process. Consideration should be made for specific product and process requirements.

Humidity Requirement
The normal operating humidity requirement.

Particulate Levels
Particulate levels are specifically defined for each room classification "at rest" and "in operation". The levels are controlled though air filtration, facility design, gowning requirements, and decontamination

Room Exhaust
Where there is a risk of active compounds being present in extracted air, filters should be fitted, preferably in the room, to prevent contamination of ductwork and the environment. The filters must be selected based on the particle size distribution of the products to be handled.

HVACS

The HVAC system must be appropriately selected using the specific design requirements as outlined above. The system must be able to provide clean, conditioned air to the specified areas to meet all of the quality requirements.

The most important precursor to HVAC design is the comprehensive definition of the function and performance required followed by the selection of an appropriate system. A poor selection can lead to unnecessarily high-energy consumption, and operational deficiencies. All-air systems rely on the movement of large quantities of air through a central air handling unit to control room conditions, as well as provide for ventilation requirements. They have the advantage of being relatively simple with most of the unit situated in one location; however, they are very space consuming. All-air systems tend to be relatively inflexible and not ideal for areas that are likely to need environmental alteration on a regular basis.

These HVAC systems are used for areas that have a lot of small zones, each with slightly different thermal loads but which requires constant ventilation. These systems can have poor energy efficiency if a lot of reheat is required. These are typically used in large manufacturing areas, and laboratories with many small rooms.

Dust Extraction and Collection
It is essential to capture dust as close as possible to the point of generation without affecting the process. In most cases dust capture should be within 100mm from the point of release. Air velocity is the key parameter in dust capture.

Pharmaceutical and chemical applications have specific collection requirements as any dust build-up in the system is likely to be of a pharmacologically active nature, sensitising, toxic and/or corrosive. It is vital to maintain transport velocities and minimise any potential for cross contamination.

A typical system should have a minimum transport velocity of 18 m/s, but this may need to be higher if heavy particles are to be collected. This velocity must be maintained throughout the system to prevent dust from dropping out in the ducts. The dust collection must be configured with the hazardous nature of the dust in mind. A clearly defined disposal procedure for the collected dust (e.g. bag-in / bag-out system for filter and dust bin) needs to be understood at the design stage. HVAC unit shall meet EN 1886 and EN 13053 requirements.

Fans
Certified performance curves are required to verify correct fan operation. Fans that may be subjected to high temperatures, humidity, corrosive fumes or other hazardous atmospheres should be constructed using non-reactive, non-corrosive, suitable and approved materials (such as epoxy painting). Whenever H_2O_2 or other disinfection application is planned, material compatibility certificates shall be supplied by the vendor.
Fans must be selected to supply the design volume, taking into account the assumption that filters are half clogged, except for the terminal filter which shall be considered to be fully clogged according to EN 13053. If the terminal filter is HEPA, clogging shall be considered according to EN 1822 and the target volume is 80% of the given maximum clogged specified value.

Filtration
Face-fitting filters shall be used in all cases, as slide-in filter elements never give a good seal. The installation must be such that the airflow pushes the filter against the seal. The face velocity across the filter section shall not exceed 2 m/s. For ventilation and air conditioning applications, two minimum filtration stages are required. For certain applications, return air filtration will be required to contain highly active materials (e.g. viruses or potent compounds). Normally, these filters should be changed from the room side. However, since those filters must be integrity tested, it is recommended to place one filter in the main return duct before the exhaust fan and design return duct network, in order to ensure tightness of the duct between the room and the filter (bag-in / bag-out filter change systems should be provided for BSL-3 areas). In case of live biological agent biocontainment, decontamination up to the filter must be proven. The grade of filter and technical solution must be selected based on the product particle size distribution and occupational exposure band (OEB) level.

HEPA filters and Dehumidification

For most HVAC applications, dehumidification is best achieved by the use of cooling coils. It should be noted that dehumidification is a very high consumer of energy and should only be used if there is a real process need. When areas are not in use, the dehumidifier should be turned off, if possible.

When room humidity must be maintained below 50% during warm weather, an absorption dryer may be necessary unless the room temperature can be increased within specification to compensate.
Normal practice is to optimise size and efficiency of the absorption dryer by first removing as much moisture from the air as possible by cooling. The design of absorption dryers is normally based on a slowly rotating desiccant wheel.

Air is passed through the wheel and dried by the desiccant coating (guidance: lithium chloride especially if the wheel is not used frequently and silica gel if used permanently and with low humidity target). It is not normally necessary to size a dryer to handle the entire air volume. Drying a proportion of air and re-mixing to achieve the desired moisture content is usually sufficient.

Air humidification may be necessary during cold weather when introducing fresh air to spaces that require humidity control. When air humidification is necessary, humidifiers should be selected on the following basis:
- direct steam injection using steam
- direct steam injection using self-generative electric or gas steam humidifier.

Humidifiers should be located before the fan and the final filter which will remove any particulate generated. At least 300 mm clearance should be allowed upstream and 1 m downstream between humidifier manifolds and coils, attenuators etc. (general recommendation to be confirmed through calculation note provided by the vendor). A single manifold or multiple manifolds in parallel may be used to meet the humidification requirements as per manufacturer's recommendations.

Sound Attenuators
Sound attenuators should be provided as necessary, to achieve the specified noise levels within occupied spaces. To minimise external noise nuisance, assessment can confirm the necessity to use acoustic media (enveloped in polyester film), that is inert and corrosion-resistant at normal operating conditions. Material quality shall be equivalent to that specified for HVAC unit or ducts.

Sound attenuators should be installed in the air handling unit or ductwork. The use of sound attenuators in the air supply and air return should be based on requirements for fresh air inlet and air exhaust, and according to external noise levels that might need to be maintained at or below the ambient site noise levels.

Dampers
The provision of sufficient dampers is essential for proper control. To minimise noise transmission into the room, these should be mounted as far as possible from the diffuser.
Carefully evaluate the space-by-space pressure control that will be used in the design. Static pressure control via hard balance or dynamic control via air terminal control units are both appropriate. Consideration should be given to the overall project size, the complexity of the facility and the project budget.

Automatic volume controllers are recommended for regulating air volume independently of supply pressure. They can be selected for constant volume, variable volume or dual duct mixing applications. Automatic low-leakage fresh air and exhaust air shutoff dampers are strongly recommended to isolate the HVAC network. Fresh air dampers shall be Class 3 minimum (maximum leakage preventing coil freezing). Whenever fumigation is performed shutoff damper shall ensure Class 4 leakage rate. Where dampers are required to provide modulating control of airflow, they must be selected to provide an appropriate level of control authority. This will normally mean a damper smaller than the duct size.

Heating and Cooling

Heating mode: Low pressure hot water (LPHW) is the preferred heating medium for HVAC applications and should be used whenever practicable. Electrical heating should be avoided due to fire risk and should be limited to low power coil and in locations where no other energies are available. Hazard operability analysis (HAZOP) must be conducted if electrical heating is being considered. Cooling mode: Chilled water is the preferred cooling medium for HVAC applications and should be used whenever practicable.

The direct expansion of refrigerant in coils is an acceptable method of cooling, particularly on small isolated plants, or where lower temperatures are needed for dehumidification or for cold room. This system, however, does not normally give close control. Direct expansion coils should only be used with extreme care on variable air volume systems (if speed driver available on compressors).

Heating Coils
The face velocity of air across heating coils should not exceed 2 m/s. Coils should be made of material suitable for applicable constraints. Drains shall be located outside the casing of the HVAC unit. Coils shall be removable.

Cooling Coils
Cooling coils have been identified as potential sources of microbial contamination; therefore, careful design is required to prevent water carryover and to ensure that drain pans do not retain water. Double tube, non-welded units are recommended. The face velocity of air across cooling coils should not exceed 2 m/s. Where necessary, stainless steel or plastic eliminator blades should be provided to prevent any moisture carryover. Where provided, these must be removable for cleaning.

Ductwork
For most applications, galvanised steel ductwork will be the most appropriate form of construction; however, stainless steel or plastic construction may be necessary where there is a higher risk of corrosion due to moisture or fumes (exhaust ducts usually). Where operating pressures above 2,000 Pa are necessary, fully welded construction is recommended. For contained ducts (e.g., exhaust duct before bag-in / bag-out filter), air tightness Class C shall be followed (EN 12237). For BSL-3, fully welded construction should be considered.

Generally ductwork should be constructed to an appropriate local standard, suitable for the maximum design pressure (positive or negative), such as those published by Sheet Metal and Air Conditioning Contractors' National Association (SMACNA) in the USA, Building and Engineering Services Association (B&ES) in the UK . Where flexible connections are proposed these must be designed for the same pressure as the ductwork. Solid ducted connections are preferred for final connections to terminal HEPA filter housings. For applications where flexible connections to diffusers are used, these should be no longer than 500 mm and nominally straight.

Special consideration must be given to fume extract ducts where these pass through fire barriers. Using fire dampers should be avoided where the loss of extraction could make a fire situation worse. An alternative design, such as the use of fire-rated ductwork, may be necessary in these cases. A thorough risk assessment must be conducted.

Environmental Monitoring

An environmental monitoring programme is required for GMP controlled areas. The purpose of such programmes is to document, define and describe parameters to be monitored, monitoring frequency and methods. Environmental monitoring is a regulatory requirement. It also demonstrates that the GMP areas are been controlled and are fit for purpose.

Grade A, B and C
- ➤ Viable and non-viable particles monitored under operational conditions
- ➤ Risk-based approach to sampling points that represent high risk/critical positions

Grade D
- ➤ Non-viable particles must be measured at-rest conditions

> Viable particles measured under operational conditions

A Building Management System (BMS) is an automated control system that is used to manage building and facilities heating, ventilation and air conditioning, security, fire protection systems and so on. It is made up of many different input / output subsystems, controller(s), server(s) and workstation(s) communicating over a control network to control, monitor, alarm and trend equipment. . BMS systems are also referred to as a Facilities Management/Monitoring System (FMS), Energy Management System (EMS), Building Automation System (BAS) or other equivalent.

Environmental Monitoring Systems (EMS) are automated control systems consisting of input / output subsystems, controller(s), server(s) and workstation(s) communicating over a control network to monitor, alarm and trend environmental critical process parameters such as temperature, humidity, differential pressure, conductivity, cooler / refrigerator status amongst others.

Contamination Control

The philosophy of containment control requires it to be applied across all inputs that make up a facility, equipment, processes, utilities and so on. Containment is primarily concerned with keeping things in by preventing product or processing agents from egressing into the surrounding atmosphere. Ensuring adequate containment protects personnel who interact with the process, equipment and systems. Aseptic processing often deals with biological agents or compounds that may be harmful to operators or technicians.

A secondary concern of containment is protection of the environment. Containment also complements efforts in contamination prevention. As with aseptic processing the risk to the patient and product must be at the forefront of activity. Risk-based approaches and tools should be used to identify potential risks and put in place adequate controls and mitigations. Any assessment should take into account all the following systems:

> Facility layout
> Drainage systems
> HVAC requirements
> Location and adequacy of utilities
> Personnel flow and procedures for entering and leaving
> Behavioural requirements of personnel in the clean room
> Flow of materials and products to prevent cross-contamination and mix-ups between products and between dirty and clean or sterile and non-sterile equipment and products
> Design to avoid cross-contamination when manufacturing live

biological agents, e.g. local exhaust air HEPA filtration, dedicated air handling units.

Material Flow

The design and layout of any manufacturing area should facilitate the effective flow of materials. This is a fundamental requirement no matter what the industry, e.g. medical devices, pharmaceuticals, bio pharmaceuticals and even non-regulated engineering companies that assemble, machine or fabricate products. However, the manufacture of medicinal products that are required to be sterile imposes a greater level of control and thought.

With regard to aseptic processing facilities, material flows do not only require efficient and effective flow of materials; the activity should also support the requirements of aseptic processing while minimising any risk of contamination. Identifying critical processing zones is a crucial step in ensuring the right building design and controls are implemented. Isolators and aseptic filling require the highest classification with strict environmental controls. Secondary packaging operations such as cartonning are often completed in areas controlled and operated to a lower classification.

Design and layout of facilities should:
- ➢ Maintain microbiological integrity of the identified critical processing zones
- ➢ Prevent or minimise contamination from outside critical processing zones
- ➢ Control the flow of materials by restricting access to trained and authorized personnel

Material Transfer

Material transfer from the outside of clean rooms to the inside is completed via material air locks or hatches. Material air locks and hatches ensure that there is clear separation between controlled clean areas and less clean areas. Many suppliers provide products that are double bagged. This provides an added level of control when transferring materials. The outer bag can be removed within the air lock thus providing a clean inner product. Material air locks also allow the sanitisation of products. Tools and other items must be clean and dirt free.

Controls that prevent personnel from the clean area and less clean area being present in the material air lock at the same time. This can be achieved by training and educating staff on the importance of contamination control. A simple visual check of the air lock to confirm it is vacant can be done in order to avoid mixing of personal from different zones. Decontamination procedures are necessary to ensure materials or tools entering the controlled area are decontaminated.

Material Air Lock Considerations:
- Interlocked doors
- Access control
- Sanitation/cleaning procedure
- Double or triple bagged products
- Dedicated trolley for air locks

Disinfection and Cleaning Agents

When materials are being transferred via an air lock, consideration must be given to the status of materials and products. As a rule, no cardboard or unnecessary paper should enter a clean room. Wooden pallets are not acceptable as they can carry dirt and microorganisms and wood cannot be sanitised due to its porous nature. Soft fabric cases often used to carry tools should also be avoided as the material can carry dirt and grease. Cleaning and disinfecting agents should be tested and approved prior to their use onsite. The choice of agents should be backed up with studies that demonstrate the effectiveness of disinfectants and cleaning agents.

Gown-Up Areas

Gowning rooms are designed in order to minimise contamination and facilitate the orderly change over from street clothes to scrubs and/or gowns. Hand washing facilities help reduce the risk of humans carrying unwanted microorganisms into the aseptic processing area. The design of the room should result in clear separation between the less clean side and the clean side. This can be achieved with a step-over segregating the two areas.

Other features of gowning rooms should include:

- Storage lockers for street clothes
- Gown and garment storage
- Body-length mirrors
- Hand washing /drying and disinfection facilities

Area Classification

Selecting a suitable classification for a room or manufacturing facility depends on several factors. Firstly, it can be said that sterile products require a more stringent set of criteria than non-sterile products.
However, there is an extensive range of products and medical devices that are sterile but are used in different ways and consist of different materials and technology. Some sterile products are single-use only and used for short term purposes and then disposed of. Other sterile products are used subcutaneously for longer periods or even require implantation.

Therefore, the design of a facility along with its HVAC specification must be appropriate to the product being manufactured. High-risk products require greater control.

The goal of facilities and HVAC systems is to minimise contamination and the associated risks. Using a "sterile versus non-sterile" rule of thumb is not adequate when classifying a room or facility. Standards including EN ISO 14644-1 and guidelines such as EU cGMP Guidelines EudraLex volume 4 Annex 1 (2008) should be consulted in order to fully understand the requirements of each ISO classification and grade of room.

ISO classifications do not specify room occupancy states but when a designation is applied, the occupancy state must be stated in the relevant documentation or procedure. The most relevant European guideline (Annex 1 of the EU cGMP Guideline) lists four classification grades and their associated particulate limits in the 'at-rest' and 'in-operation' conditions. In general, for the sterile and non-sterile products, similar classes are applied, but in non-sterile production the producer could assign their classes, having similar particulate concentration, temperature, pressure etc. but lower air-change rate could be used.

Types of Contamination:

- cross contamination (of a product/material with another product/material)
- non-microbial particulate contamination (non-viable particles)
- biological/microbiological contamination (viable particles/micro-organisms)

Factors Influencing Contamination Cleanliness Levels in the Manufacturing Processes:
- process
- air cleanliness

- personnel hygiene and clothing
- work practices
- material design (material of construction, surface finishes, room finishes, equipment, open system/enclosed system, utensils, etc.)material cleanliness

Environmental Grade A (Aseptic)

Grade A is reserved for critical processes in manufacturing sterile products, product components or product contact. This is generally achieved using isolator technology which maintains a barrier to the background environment or surrounding room.

Grade A **Operations** include:

- Aseptic processing of sterile ingredients
- Filling of sterile products not for terminal sterilisation
- Stopper insertion
- Crimp capping

Environmental Grade B

Grade B is used for supportive work for aseptic processing corresponding to ISO 14644 (Part 1) Class 5 ("at-rest") and Class 7 (when "in-operation"). Grade B areas typically serve as the background environment of Grade A areas for aseptic processing.

Environmental Grade C

Suitable for non-critical processing steps, Grade C corresponds to ISO 14644 Part 1 Class 7 ("at-rest") and Class 8 ("in-operation"). Grade C operations include:

- Clean side of material air locks and gowning rooms
- Filling of products that are to be terminally sterilised

Environmental Grade D

Grade D at least corresponds to ISO 14644 Part 1 Class 8 ("at-rest" / no definition for "in-operation").

- Clean section of material air locks and final compartments of gowning rooms

- Dispensing of raw materials and excipients and preparation of solutions for sterile products to be sterile filtered and terminally sterilised
- Background environment for transfer and crimp capping of stoppered containers with sterile products

Compliance Tests

Test	Requirements
Particle count test	Test covers verification of cleanliness. Dust particle counts to be carried out and result printed. The number of readings and positions of tests should be defined in accordance with ISO 14644-1 Annex B5.
Air pressure difference	This test is used to verify non cross-contamination. Log of pressure differential readings to be produced or critical plants should be logged daily, preferably continuously. A 15 Pa pressure differential between different zones is recommended. Refer to ISO 14644-3 Annex B5.
Airflow volume	To verify air change rates. Airflow readings for supply air and return air grilles to be measured and air change rates to be calculated. Refer to ISO 14644-3 Annex B13.
Airflow velocity	To verify unidirectional flow or containment conditions. Air velocities for containment systems and unidirectional flow protection systems to be measured. Refer to ISO 14644-3 Annex B4.
Filter leakage tests	To verify filter integrity. Filter penetration tests to be carried out by a competent person to demonstrate filter media, filter seal and filter frame integrity. Only required on HEPA filters. Refer to ISO 14644-3 Annex B6.
Containment leakage	To verify absence of cross-contamination. Demonstrate that contaminant is maintained within a room by means of: • airflow direction smoke tests • room air pressures. Refer to ISO 14644-3 Annex B4.

Recovery	To verify clean-up time. Test to establish time that a clean room takes to recover from a contaminated condition to the specified clean room condition. Should not take more than 15 minutes. Refer to ISO 14644-3 Annex B13.
Airflow visualisation	To verify required airflow patterns. Tests to demonstrate air flows: • from clean to dirty areas • do not cause cross-contamination • uniformly from unidirectional airflow units Demonstrated by actual or video-taped smoke tests. Refer to ISO 14644-3 Annex B7.

Further reading

ISO 14644-1: International Organisation For Standardisation – Cleanrooms and Associated Controlled Environments. Part 1: Classification of Air Cleanliness.
ISO 14644-3: International Organisation For Standardisation – Cleanrooms and Associated Controlled Environments. Part 3: Test Methods.
ISO 14644-4: International Organisation For Standardisation Cleanrooms and Associated Controlled Environments: Part 4: Design, Construction and Start-Up.
EudraLex, Vol 4, Annex 1: EU Guide to Good Manufacturing Practice (EU GGMP) Governing Medicinal Products for Human and Veterinary Use, Annex 1 – Manufacture of Sterile Medicinal Products.
EN 1822:2009: European Standard For HEPA Filter Classification.
US FDA CFR 211: Code of Federal Regulations Food and Drug Administration Title 21 Part 211 – Current Good Manufacturing Practice for Finished Pharmaceuticals – Section 211.46 Ventilation, Air Filtration, Air Heating and Cooling.
ICH Q7: International Conference on Harmonisation - Good Manufacturing Practice Guide for active Pharmaceutical Ingredients – Section 4.21 and 4.22 – Utilities.
US FDA: Food and Drug Administration - Guidance for Industry "Sterile Drug Products Produced by Aseptic Processing – Current Good Manufacturing Practice".

Utilities

Introduction

The term "clean utilities" in the life science industry refers to utilities that have to fulfil regulatory requirements. The most common utility is water, which can be supplied in different pharmaceutical grades of purity. Purified water (PW or PUW), highly purified water (HPW) and water for injection (WFI) are the most common. Water quality specifications can be found in the pharmacopeias, e.g. the US Pharmacopeia. Other clean utilities can also include clean compressed air, clean gases (e.g. nitrogen, argon and oxygen), and clean steam.

Compressed Air

Compressed air is used for valve actuation, instrument air and process air to name but a few applications. Only the point-of-use filtration and the gas quality instrumentation should be classified as level 1. When flow or pressure is a CPP, the measurement/monitoring should be performed by the system into which the gas is flowing. Additionally, the CQAs and CPPs should be routinely monitored through the calibrated monitoring system. For compressed air, the potential CPPs are listed below. For the physical system being evaluated, the use and the application of the compressed air will determine which (if not all) CPPs are needed to ensure the system produces product of the desired quality.

- Hydrocarbons
- Moisture
- Particulates
- Temperature

It is important that each point of use has appropriate sterile filters in place. If the filter is not placed directly at the point of use, control and counter measures should be implemented to address any risk of contamination downstream of the filter. Compressed air for bio-pharmaceutical use must be generated using oil-free compressors with appropriate temperature controls in place.

Water Systems

Water supply and the associated water systems in biotechnology and pharmaceuticals are vital components of the manufacturing process. They are used to clean equipment and vessels, to cool or heat processing pipes and systems, and in many circumstances certain grades of water are components of the finished product (e.g. water-for-injection). Various grades of water service a particular purpose. Some common types include:

- Potable water
- Soft water
- Purified water
- Water-for injection

Water used in-process and in-cleaning should be pure and free from microbial and chemical impurities. As the water gets easily contaminated by environmental conditions, diligence in the design is essential. Typically water systems are supplied on a continuous loop with recirculation.

CPPs typical for a water system include:
- Pressure
- pH
- Conductivity
- Level
- TOC
- Flow
- Temperature
- Resistivity

Water-for-Injection

The use of WFI is twofold. Firstly, it can be used for critical processing steps such as washing and rinsing. It can also be used in injectable products. WFI is a key raw material for sterile intravenous and intradermal products. WFI is produced by Multi Column Distillation Plant (MCDP), and must meet the microbial requirements of regulated bodies.

Clean-in-Place (CIP) / Sterilise-in-Place (SIP) System

The cleaning of equipment, vessels and process piping is a critical activity. Any residue from a previous production batch needs to be removed in order to avoid cross contamination. CIP and SIP skids are often utilised to allow efficient switchover between batches and/or products.

Clean steam

Pure Steam is used in pharma and biotech for sterile application, autoclave sterilisation etc. Distribution piping of clean steam is a critical aspect. Improper sizing of pipes may impact the production process and lead to loss of time during sterilisation.

Clean steam, also referred to as "pure steam", and gases used in manufacturing operations must be of a quality suitable for their intended purpose. The intended use of clean steam and gases must be understood in order to determine any risks to the patient or product. For example, gases that end up being part of the product must fulfil the regulatory requirements. Preventative maintenance and on-going monitoring must be implemented for clean steam systems.

- Routine inspection and maintenance
- Frequency of filter change
- Frequency of the sterilisation for the gas distribution system, if applicable
- Frequency for integrity testing of the sterile filter

Water systems for purified water, de-ionised water and water-for-injection (WFI) must provide a consistent and reproducible output. Where there is moisture, there is always a risk of microbial contamination. Therefore, the design of water systems should mitigate against such risks. Good engineering practices such as using circulation loops, no dead legs and polished surface finishes all work to provide an effective and safe system. The design should also take into account ease of sampling at the point of use. The removal of endotoxins is a requirement for WFI. On-going sampling to monitor the quality of water is particularly important where water systems are concerned. Procedures should be in place to ensure effective monitoring and testing is maintained. Action limits and acceptance criteria should be clearly documented in approved SOPs or equivalent. Failure to meet limits or acceptance criteria should initiate an investigation. The potential CPPs are listed below for clean steam systems:

- Conductivity
- Flow
- Level
- Pressure
- Resistivity
- Temperature

Heating, ventilation and air-conditioning (HVAC) plays an important role in ensuring the manufacture of quality products. Furthermore, HVAC systems also provide comfortable conditions for operators based in the manufacturing environment. HVAC system design influences the layout of airlock positions and doorways. In turn, airlocks, entrances and exits have an effect on room pressure differential cascades and cross-contamination control.

The prevention of contamination and cross-contamination is an essential design consideration of the HVAC system. In view of these critical aspects, the design of the HVAC system should be considered at the concept design stage of a manufacturing plant.

Temperature, relative humidity (RH) and ventilation should not adversely affect the quality of products during their manufacture and storage, or the proper functioning of equipment. CPPs for HVAC systems include:
- Temperature
- Humidity
- Particle count (viable and non-viable)
- HEPA filter certification/leak test/air flow rates
- Room differential pressures

Sterile Manufacturing

Introduction

Sterile manufacturing operations depends on several factors including the right design and operation of facilities, utilities and equipment. Sterility assurance must be demonstrated to be in control within a manufacturing setting. This is achieved by:

- Qualification and validation of the processes, facilities, utilities, equipment, cleaning methods and sterilisation operations
- Qualified personnel for aseptic handling in conventional clean rooms or by barrier systems
- Control of critical aspects and critical parameters via the application of change management, change control and a suitable quality management system
- Environmental monitoring
- Routine Maintenance
- Analytical method validation

The impact of contaminated injectable products can result in serious illness or death to patients. Many injectable treatments sustain life and bio-chemical processes or genetic conditions. While there is always residual risks or acceptable risks, it is important to mitigate against any risks throughout the manufacturing process. Furthermore, a risk-based approach to operations and in particular changes to the process must be maintained throughout the life cycle of a product. Contamination can be caused by particles or microbes.
Where appropriate and technically permissible terminal sterilisation is the preferred point of sterilisation. Terminal sterilisation is when the final sealed product in its container is sterilised at the end of the process.

Unit Operations

Bioprocesses treat raw materials and generate useful products. Unit operations are the individual steps in the process that modify materials and their properties at each step of the process. Each unit operation comes together to create a complete process. The term unit operation usually refers to processes that cause physical modifications to materials such as a change in phase or component concentration. Chemical or biochemical changes are the subject of reaction engineering.

Bioreactor Engineering

The design and manufacture of bioreactors is yet again an area within bioprocessing that depends on scientific and engineering expertise. It should be pointed out that there is no standard design procedure for the design of reactors. However, knowledge of bioprocess reactions and kinetics is a key element. Other knowledge such as mixing, mass transfer and heat transfer also contribute to the design process. Key aspects of bioreactor design include:

Reactor size: What is the capacity of the reactor? This is generally driven by the expected production volumes.
Reactor configuration: Is the reactor air driven, stirred, agitated etc.?
Operating configuration: Is it a continuous operation or a batch driven operation?
Process Requirements: Refer to the required operating temperatures, pH that needs to be maintained in the vessel.

Stirred Tank

A conventional tank involves mixing and bubble dispersion done via mechanical agitation. This requires a high energy input per unit volume. Headspace is an important consideration when filling tanks. Typically, only between 60% to 80% of the tank volume is used. This headspace is important especially if foaming of the broth occurs. Some tanks are designed to take account of foaming issues with the addition of a foam. Chemical means of reducing or preventing foam formation can also be employed. However, these chemicals can impact the process (reduction in rate of oxygen transfer). Temperature modulation is typically controlled using coils.

Bubble Column Bioreactor

A bubble column is a type of bioreactor. Bubble columns offer an alternative to stirred reactors, having no mechanical means of stirring. Mixing and aeration is done by gas sparging by the use of a gas sparger placed at the bottom end of the vessel. This type of reactor requires a lot less energy to mix compared to mechanical stirring.

Airlift Reactors

Airlift reactors are similar to bubble columns as neither require mechanical mixing. A key difference between bubble columns and airlifts is that the air is channelled through a riser in the airlift, which allows more control of the bubble patterns. Airlift reactors can be categorised into either internal loop or external loop configurations.

Aseptic Operation

With the exception of food and beverage fermentations, cultures used in the treatment of medical conditions frequently require sterile conditions.
This is especially important for slow growing cultures that can be quickly compromised by unwanted contaminates. Typically, up to 5% of fermentations in industrial settings are lost as a result of failings in sterilisation. Slow growing cells would have a higher rate of contamination due to sterility issues. Antibiotics by their nature have a higher resilience to this type of loss.
Industrial fermenters are designed to allow in-place steam sterilisation under pressure.

For effective steam sterilisation, the vessel must be fully purged of air. Dead legs, stagnant areas or crevices should be avoided during the design phase as these can be a point of microbial contamination. Polished welded joints with a high surface finish are desired.

Valves

Valves control the introduction of liquids to the vessel and their removal when required. Valves therefore, can be a potential entry point for contaminants. Traditional gate and globe valves do not suffice for aseptic operations.
Pinch and diaphragm type valves are more commonly used as they do not contain any dead spaces within their assembly. The closing mechanics also provide isolation from the liquid or product contents.

Materials of Construction (MOC)

Fermenters are made of materials that are suited to the use of steam sterilisation techniques and regular cleaning. These materials can be classed as both non-reactive and non-absorptive surfaces. Most large-scale reactors are made of high-grade stainless steel. Cheaper classifications of stainless steel can be used for jacketing and other non-product contact areas.

All interior product contact surfaces should be polished to a "mirror" finish. Welds also need to be finished in a similar manner. Electro polishing provides a better-quality surface finish than mechanical polishing.

As with any chemical reaction, factors such as temperature, pH and oxygen concentration can impact the performance and yield. To ensure the optimum conditions are maintained, it is important to monitor and control such parameters and factors. By far the most common these days is automatic control of systems and equipment with automatic feedback and adjustment.

Filtration

Basic filter design involves solids being retained in the filter cloth, while the liquid passes through the cloth/membrane. However, the liquid filtrate that passes through typically contains a small portion of solids. It should be noted that large scale filtration is expensive and difficult to perform under sterile conditions.

Microfiltration

Microfiltration uses microporous membranes to recover cells. Unlike filtration, microfiltration generally does not require preconditioning (heating or addition of agents to reduce viscosity). Microfiltration allows cell recovery of typically 100%, so it is therefore very efficient. Microfiltration can also be done under sterile conditions. It is also typically less expensive than filtration and centrifugation.

Membrane Filtration

Membrane filtration is another type of unit operation that is used in downstream processing. It can be applied in order to separate, concentrate or purify a product. Applications include:

- Cell removal
- Cell debris removal
- Desalting
- Removal of viruses
- Recovery of precipitates

Membrane filtration has a number of advantages compared to other unit operations used to concentrate products:
- Low process energy requirements
- Membrane filtration can be done aseptically
- Does not need harsh chemicals

Membrane filtration can be categorised according to the size of the particles that are retained by the membrane:

Microfiltration: used to remove particulate such as cells and cell debris ranging in size from 0.2 to 10μm from broths. Typical membranes have a nominal pore size diameter of 0.05 to 5μm.

Ultrafiltration: Membranes for ultrafiltration have pores typically of a nominal size between 0.001μm to 0.1μm.

Raw Materials

Raw materials and components used in the manufacturing process should be properly sourced and approved through a supplier quality programme. This ensures that the vendor or supplier of raw materials has the necessary regulatory status and quality controls in place. A robust supplier approval process ensures materials are provided consistently to pre-approved specifications. Raw materials for sterile products must be tested for their bioburden and when necessary for bacterial endotoxin levels to determine acceptability of their use.

Filling Operations

Suspensions and solutions that are filled in glassware such as vials provide lifesaving and sustaining medical treatments for millions of patients worldwide. When the product reaches the filling unit operation, it has been through many unit operations. The product and components must be sterile at this point. Transfer of product to individual vials or containers may be facilitated by employing piston valves, pressure control and peristaltic pumps.

Once the required quantity of solution or suspension has been filled, the next unit operation required is container closure achieved by the insertion or application of a stopper or cap. Key consideration for filling and closing operations include:
- Design and function of filler heads
- Design and function of filler needles
- Fill accuracy and fill weight

The filling of Biotechnology Derived Products (BDP) into ampules or glass vials presents similar problems as with the processing of conventional products. Attempting to develop a site, prove clinical effectiveness and safety, as well as the validation of sterile operations, equipment, processes and systems often necessitates a lengthy process to achieve success for a start-up BDP facility.

The batch size initially produced by a BDP is likely to be small. Because of the small batch size, filling lines may not be as automated as for other products typically filled in larger quantities. Thus, there is more involvement of people filling these products. This can present more chances of contamination meaning any operation or involvement must be controlled and monitored.

Problems that have been identified during filling include inadequate attire, deficient environmental monitoring programmes, hand-stoppering of vials, particularly those that are to be lyophilised and failure to validate some of the basic sterilisation processes. Because of the active involvement of people in filling and aseptic manipulations, the number of persons involved in these operations should be minimised, and an environmental programme should include an evaluation of microbiological samples taken from people working in aseptic processing areas.

Another concern about product stability is the use of inert gas to displace oxygen during both the processing and filling of the solution. As with other products that may be sensitive to oxidation, limits for dissolved oxygen levels for the solution should be established. Likewise, validation of the filling operation should include parameters such as line speed and location of filling syringes with respect to closure, to ensure minimal exposure to air (oxygen) for oxygen-sensitive products. In the absence of inert gas displacement, the manufacturer should be able to demonstrate that the product is not affected by oxygen.

Typically, vials to be lyophilised are partially stoppered by machine. However, some filling lines have been observed that utilise an operator to place each stopper on top of the vial by hand. The concern is the immediate avenue of contamination offered by the operator. The observation of operators and active review of filling operations should be performed. Another major concern with the filling operation of a lyophilised product is assurance of fill volumes. A low fill would represent a sub-potency in the vial. Unlike a powder or liquid fill, a low fill would not be readily apparent after lyophilisation, particularly for a product where the active ingredient may be only a milligram. Because of the clinical significance, sub-potency in a vial can potentially be a very serious situation. A common method of filling vials consists of a two-step filling process. Generally, the first step fills up to 90% of the vial, with the second more accurately filling the remaining amount. The following parameters must be maintained to achieve the same fill volumes at each filling cycle:

- Viscosity of the product
- Product temperature
- Pressure in the dosing vessel
- Level in the dosing vessel
- Needle/filling head properties
- Properties of the hose material

Container Closure Integrity

Upstream processes need to take into account the many requirements that aim to produce products that are safe and effective for patients. Operations such as dispensing and compounding apply GMP principles from the very beginning of the manufacturing process. When the product has been manufactured and is ready to be filled and closed, so too the container closure methods must ensure that sterility and integrity of the product is preserved. Therefore, sterile product container closure systems (or closing systems) must be designed, qualified, and controlled in accordance with international and local regulatory requirements and GMP guidance.

Regulatory Requirements

FDA 21 CFR Part 600. PART 600 -- BIOLOGICAL PRODUCTS: GENERAL Subpart B--Establishment Standards, h)
h) Containers and closures.

"All final containers and closures shall be made of material that will not hasten the deterioration of the product or otherwise render it less suitable for the intended use. All final containers and closures shall be clean and free of surface solids, leachable contaminants and other materials that will hasten the deterioration of the product or otherwise render it less suitable for the intended use. After filling, sealing shall be performed in a manner that will maintain the integrity of the product during the dating period. In addition, final containers and closures for products intended for use by injection shall be sterile and free from pyrogens. Except as otherwise provided in the regulations of this subchapter, final containers for products intended for use by injection shall be colourless and sufficiently transparent to permit visual examination of the contents under normal light. As soon as possible after filling final containers shall be labelled as prescribed in 610.60 et seq. of this chapter, except that final containers may be stored without such prescribed labelling provided they are stored in a sealed receptacle labelled both inside and outside with at least the name of the product, the lot number, and the filling identification."

Quality Requirements

Product containers and closure systems must be capable of being sterilised and depyrogenated before the product is filled. A simple example would be glass vial and stopper components used for various injectable products undergoing sterilisation as part of the process (depyrogenation and sterilisation tunnels for glass components, and stopper sterilisation using vessels and moist heat).

During the product development stage, the type of container closure system must be developed based on the intended use, physical and chemical requirements of the medicine, storage requirements, delivery methods and shelf life. A detailed testing strategy should be developed as early as possible to test for the suitability of the container components. Test strategies can be best developed using a risk-based approach along with knowledge and experience of personnel. In addition to the components used and the size and shape of components, engineering studies are also required in order to define the critical parameters to be used during container closing operations. Depending on the methods of closure, some critical parameters may include:

- Crimping force (in crimp caps are used)
- Closure torque
- Stopper position
- Stopper force
- Closure

The parameter selection must ensure the integrity of the container closure system is not compromised. Principles of quality management must be applied to container closure systems when validated and should address the following:

- Approved container closure components
- Control srategy for critical parameters
- Finished product release testing
- Changes to the container closure system managed under formal change control procedures

Testing

Integrity testing is the most critical test with regard to closed containers. Depending on the container closure format, the followings tests may be used:

- Dye bath test
- Vacuum test
- Headspace analysis
- Electronic spark test
- Bubble testing
- Leak testing
- Pressure decay
- 100% visual inspection
- Presence of stopper or closure cap

- Presence of tip/cap
- Presence of cracks, sealing issues or evidence of crimping deficiencies e.g. crimp height.

<u>Electronic Spark Test</u>
This test allows the detection of very small pinholes or cracks that may cause leaks or effect the container integrity. It is used to assess ampoules, vials and glass cartridges. By placing high-voltage through the item and measuring impedance, any cracks or closure issues can be identified in samples.

<u>Vacuum Test</u>
This is completed by applying a vacuum for a defined time and then allowing it to reach ambient pressure. This is a common test used, however it can lack sensitivity.

<u>Head Space Analysis</u>
For containers that are filled using nitrogen or some other inert gases such as argon, head space analysis provides an accurate and repeatable test method. The O2 levels can be detected by the increase in oxygen content. A non-destructive way of testing for head space analysis is to use laser measurement.

<u>Aseptic Process Simulation</u>

Validation of aseptic processing for products must include simulating the process using aseptic process simulations. For simulations of final product filling, the number of containers filled should be representative of the projected batch size and be sufficient to enable a valid evaluation, including all routine operator interventions.

Isolator Barrier Systems
An isolator is a complex barrier system designed to support aseptic processing and manufacturing. The supplied air to such systems is generally supplied through a microbially-retentive filtration system. High efficiency particulate air (HEPA) filters are capable of removing particles as small as 0.3μm making them an integral part of isolator technology.

HEPA filters should be capable of achieving Grade A (ISO Class 4.8) at-rest and in-operation. Some exceptions are permitted, such as powder filling, however, risk assessments should mitigate risk to patients. The isolator is a sealed enclosure where there is no direct opening to the external environment or room. Transfer of materials or utensils is done in a controlled manner using a decontaminated interface.

Isolator Interfaces

Depending on the design considerations and individual vendor designs, isolators can have a number of operation interfaces. The term "interface" refers to the ability of an operator or process technician to interact with the machine. The primary method of intervention utilizes glove systems. Four part glove systems consisting of a gauntlet, glove, cuff-ring and sleeve. When used properly and by trained personnel, glove systems support critical line interventions required during aseptic processing and manufacturing.

Furthermore, isolators may also be designed in combination with smaller enclosures associated with them to allow the continuous ingress of materials through the smaller isolator into a main isolator.

Classification of Isolator Rooms

The surrounding room of an isolator should have limited access to staff (ensuring only the presence of authorised personnel), adequate space around the isolator and temperature/humidity under control for the effective utilisation of decontamination technologies (e.g. vapour phase hydrogen peroxide systems).

Regulatory authorities require background environments of aseptic production isolators to be classified at minimum in zone (Grade) D (ISO 8 at-rest). However, there is a general consensus that sterility testing isolators need not be placed in a classified clean room, but it is important that such isolator surrounding rooms impose restricted access.

Isolator Decontamination

The purpose of bio-decontamination is to remove viable bioburden on exposed surfaces inside the isolator; a decontamination process should be performed using sporicidal chemical agents associated with decontamination equipment such as gas/vapour phase decontamination systems using hydrogen peroxide (e.g. VHP) or the equivalent. A decontamination cycle is an automated machine cycle that is controlled and monitored during each stage of the cycle. Cycles can be divided into four stages:
-Dehumidification
-Conditioning
-Decontamination
-Aeration

Dehumidification: The dehumidification stage (also known as pre-conditioning) is designed to ensure that the isolator enclosure has a predefined humidity value (< 20 % RH) to ensure a proper concentration of decontaminating agent.

Conditioning: Depending on the complexity of the system, at a minimum, the isolator must have a tightly controlled temperature range, positive pressure and air velocity control. During this initial stage, the isolator doors and ports must be closed and sealed. Any defects in the barrier system should result in an alarm and abort the cycle. During conditioning, an automated leak test should be initiated to detect any breaks in the barrier system (e.g. defective gloves or seals). Heating of VHP delivery pipework also occurs. The conditioning stage is when the decontaminating agent shall reach the minimum concentration required to achieve the desired microbial reduction.

Decontamination: At this stage the VHP is maintained in the isolator according to the dosing rate contained in the recipe or cycle settings. The time and total amount of VHP must result in a kill in BIs placed within the isolator. Generally a 6 log reduction is required for a cycle to be deemed a success.

Aeration: During the aeration stage the amount of residual decontaminating agent must fall to safe levels. (< 1ppm). This is done by blowing the hydrogen peroxide carrying air out of the barrier system using fresh air.

Recommended Critical Process Parameters	Typical Units
Amount of H_2O_2 during conditioning	(g)
Dosing rate (conditioning)	(g/min)
Time for conditioning	(mins)
Amount of H_2O_2 during decontamination	(g)
Dosing rate decontamination	(g/min)
Time for decontamination	(mins)
Aeration time	(mins)

Decontamination Agents

Decontamination of isolators is achieved by the supply of gaseous sporicidal agents. These agents must be capable of killing both bacterial endospores and fungal spores.

The system typically turns liquid agents into a gaseous vapour. The decontamination agent typically used in industry is hydrogen peroxide. Other agents include formaldehyde, peracetic acid and chlorine dioxide. The rationale for selecting a particular agent should be based on technical data, sporicidal efficacy and the materials and products that come into contact with such agents. Often the starting point when selecting an agent is the manufacturer's recommendations.

Manufacturers of equipment trains are best positioned to understand interactions with seals and surfaces etc. In many cases, the equipment is designed with a particular type of decontamination agent in mind. Another source of information is the datasheets provided by the agent manufacturers. Datasheets also give an insight into the suitability of a chemical based on its purity, concentration and safety.

The below factors should be considered with regards to biodecontamination:

> Ensure as much surface area as possible of components are exposed.
> Minimise loads in order to limit the bioburden levels prior to the cycle starting.
> For filling and closing machines, design automation to ensure parts are moving during the cycle to facilitate exposure to the agent.
> Ensure all areas are dry and free of foreign objects and debris.

Containment bioreactor systems designed for recombinant microorganisms require not only that a pure culture is maintained, but also that the culture be contained within the systems. Both GLSP and biosafety levels are detailed in this section.

A GLSP (Good Large-Scale Practice) level of physical containment is recommended for large-scale research of production involving viable, non-pathogenic and non-toxigenic recombinant strains derived from host organisms that have an extended history or safe large-scale use.

The GLSP level of physical containment is recommended for organisms such as those that have built-in environmental limitations that permit optimum growth in the large scale setting but limited survival without adverse consequences in the environment.

BL1-LS
A BL1-LS (Biosafety Level 1 - Large-Scale) level of physical containment is recommended for large-scale research or production of viable organisms containing recombinant DNA molecules that require BL1 containment at the laboratory scale.
BL2-LS

A BL2-LS (Biosafety Level 2 - Large-Scale) level of physical containment is required for large-scale research or production of viable organisms containing recombinant DNA molecules that require BL2 containment at the laboratory scale.

BL3-LS

A BL3-LS (Biosafety Level 3 - Large-Scale) level of physical containment is required for large-scale research or production of viable organisms containing recombinant DNA molecules that require BL3 containment at the laboratory scale.

No provisions are made at this time for large-scale research or production of viable organisms containing recombinant DNA molecules that require BL4 containment at the laboratory scale.

Steam Sterilisers

Autoclaves or steam sterilisers are used to sterilise items such as tools, fixtures and utensils used in aseptic processing. Modern systems are designed to fulfil the requirements of FDA and EU regulatory requirements. DIN 58950/58951 is a standard in which many manufacturers design and build steam sterilisers to fulfil the requirements set out in the document. Conformance to this standard ensures autoclaves comply with the FDA and GMP directives. Industrial steam steriliser systems used in biotechnology companies comprise the following main components:

- Pressure container for sterilisation
- Vacuum pump
- PLC controller
- Human Machine Interface (HMI)
- Cycle software

The sterilisation process can be divided into three distinct stages:

- **Pre-treatment Stage:** during this stage the autoclave begins to heat up and the air in the chamber is replaced by a mixture of steam and air.
- **Sterilisation Stage:** the purpose of this stage is to kill any harmful microbes by using steam sterilisation. The temperature and pressure of the chamber is held at predefined settings for a specific period of time.
- **After-treatment Stage:** cooling, decompression and drying occurs in this stage of the cycle.

The steriliser can be loaded with the help of a loading trolley manufactured with suitable materials or an automatic loading and unloading system. The steriliser can alternatively be equipped with trays for accommodating the goods to be sterilised.

Air Leakage Test (Vacuum) Test

The purpose of air leakage testing is to verify that the chamber is vacuum-tight and can maintain the vacuum over a period of time. To avoid loose interpretations, a formal definition of vacuum-tight should be documented. The British standard EN 285+A2 "Sterilisation. Steam sterilisers. Large sterilisers" provides definitions, guidance and a framework for testing steam sterilisers. The air leakage test should result in the chamber maintaining a predetermined pressure over a set period of time e.g. ten minutes.

Steam Penetration (Bowie Dick) Test
Steam penetration is tested using a Bowie Dick test kit. To verify the consistency of the process, this is typically done three times for a recipe or cycle.

Pressure Leak Testing

This test is used to ensure the chamber does not leak. During the course of the test, the pressure is trended. The pressure drop over the test must be within specification. For example, the pressure decrease should be less than 100mbar during the course of the cycle (e.g. 10 minutes).

Depyrogenation

Depyrogenation is a thermal process that involves the removal of pyrogens from components (e.g. vials or containers) that are used for injectable pharmaceuticals and biopharmaceuticals. A pyrogen is defined as any substance that can cause a fever. Bacterial pyrogens include endotoxins. Later on we shall see that endotoxins are used to challenge depyrogenation tunnels. Depyrogenation tunnel design varies depending on the manufacturer, however, they usually consist of the following components:

> - Infeed and preheating
> - Heating zone
> - Cooling zone
> - Outfeed and transport to next unit operation
> - Automatic emptying

Pyrogens

Pyrogens are fever inducing proteins of low-molecular-weight proteins. Pyrogens of external origin are referred to as exogenous pyrogens. Modern injection and delivery systems are largely safe, yet adverse reactions are still reported. If a treatment or medication administrated via hypodermic needle is contaminated with toxins such as pyrogens fever can be induced which can lead to some death in some cases. It was known in the latter part of the 19th century that some parenteral solutions caused a marked rise in body temperature. The fever producing agents were not known, and hence described in general terms such as "injection fever," "distilled water fever," and "saline fever". Bacterial pyrogens are responsible for many of those early fevers and for many of the other biological effects described incidental to parenteral therapies. The route of administration of a drug allows a pyrogen, if present, to bypass the bodies primary defences. The host's response is mediated through the leukocytes (white blood corpuscles) which in turn release their own kind of pyrogen (endogenous pyrogen) and this in turn initiates a fever like response and other biological reactions.

Bacterial Toxins

There are two general kinds of bacterial toxins: (1) endotoxins and (2) exotoxins. Endotoxins can be extracted from a wide variety of gram-negative bacteria. The term "endotoxin" is usually interchangeable with the term "pyrogen" although not all pyrogens. Higher doses of endotoxin are required to produce a lethal effect in the experimental animal than are required for exotoxins. The effects produced by endotoxins on the host are systemic such as fever and general body reactions, rather than strictly neurological effects, as is the case with most exotoxins. Endotoxins are found in the gram-negative bacteria mostly and are obtained subsequent to the death and autolysis of the cells. The endotoxins are extracted from and associated with the cell structure (cell wall). Good examples of pyrogen producing bacteria are S. typhosa, E. coli, and Ps. aeruginosa.

Exotoxins are produced during the growth phase of certain kinds of bacteria and are liberated into the medium or tissue. Exotoxins are protein in nature and their reactions are specific. For example, Clostridium botulinum produces an exotoxin of unusual potency which affects only neurological tissue. Other well-known examples of exotoxins are tetanus toxin, Shiga toxin, and diphtheria toxin.

Properties of Pyrogens

Pyrogens are:

- ➢ Known to consist biochemically of a lipid-polysaccharide-peptide substance
- ➢ Heat stable at the temperature of boiling water
- ➢ Demonstrate a low order of immune response
- ➢ Produced from persistent gram-negative bacteraemia which could have a 50% mortality rate

Bactericidal procedures such as heating, filtration, or adsorption techniques do not eliminate pyrogens from parenteral solutions. All ingredients must be kept pyrogen-free in the first place. For this assurance, the manufacturer carries out comprehensive pyrogen screening tests on all parenteral drug ingredients and sees to their proper storage prior to use. Ideally, the manufacturer recognises the critical steps in the manufacturing operations that could allow growth of pyrogen producing bacteria and monitors these areas routinely. For example, the water in the holding tanks would be tested for pyrogens and the manufacturer would insist on minimum holding times so that only pyrogen-free water is used. Pyrogen-free water, as "water-for-injection" outlined in the USP, is the heart of the parenteral industry.

Pyrogen Assay - Limulus Amoebocyte Lysate

Many laboratories conduct pyrogen assays by means of the limulus amoebocyte lysate (LAL) test method. The LAL method is useful especially for screening products that are impractical to test by the rabbit method. Products best tested for endotoxins by LAL techniques are: radiopharmaceuticals, anaesthetics, and many biologicals. Essentially, the LAL method reacts hemolymph (blood) from a horseshoe crab (limulus polyphemus) with an endotoxin to form a gel. The quantity of endotoxin that gels is determined from dilution techniques comparing gel formation of a test sample to that of a reference pyrogen, or from spectrophotometric methods comparing the opacity of gel formation of a test sample to that opacity of a reference pyrogen. The LAL test is considered to be specific for the presence of endotoxins and is at least a hundred times more sensitive than the rabbit test. Even picogram quantities of endotoxins can be shown by the LAL method. Although LAL is a relatively new pyrogen testing method, it has produced a wide variety of polysaccharide derivatives that give positive limulus test results and also show fever activity. It is also a fact that some substances interfere with the LAL test even when pyrogens are present.

Some firms use the LAL test for screening pyrogens in raw materials and follow up with pyrogen testing on the final product by means of the USP rabbit assay. The LAL test for pyrogens in drugs requires an amendment to the NDA on an individual product basis. LAL test reagents are licensed by the Bureau of Biologics. For devices, a firm must have its protocol approved by the Director Bureau of Medical Devices, before it can substitute the LAL assay for the rabbit. What is certain is that pyrogens remain a potential source of danger with the use of parenteral therapy.

Endotoxins and Depyrogenation

Endotoxins are used to challenge the effectiveness and consistency of depyrogenation tunnels. Endotoxin challenge vials must be processed through a depyrogenation process that must demonstrate a ≥3 log reduction in endotoxin. Typically, endotoxin challenge vials are placed in close proximity to thermocouples. Using this approach, the temperature profile of the position can be obtained during a cycle. Endotoxin challenge testing is often done during SAT and process development.

Endotoxin challenge testing is typically a requirement of validation, however, no commercial product can be used during a depyrogenation tunnel performance qualification using endotoxins as the product would be potentially contaminated with endotoxins. Therefore, performance validation of depyrogenation processes results in the discarding of the vials or ampules. Depyrogenation tunnels generate tremendous amounts of heat and can operate up to temperatures as high as 320°C (depending on design and operational constraints). Air handling is also a key function of the depyrogenation tunnel. Tunnels should not allow non-sterile air from the room into the sterile air inside tunnel zones. This is done through the use of HEPA filters and an overpressure cascade approach of the tunnel compared to the surrounding room or environment. Air flow must be laminar in nature to ensure the tunnel can maintain the correct pressures and temperatures. Most tunnels are divided into two sections:

  -Hot zone (depyrogenation)
  -Cool zone (sterilisation/cooling)

The depyrogenation section typically operates at higher temperatures in excess of 270°C which is the recognised depyrogenation temperature. Depending on the technical specification of the components, set-points of 290°C, 300°C or 320°C can be used. Components move slowly through the depyrogenation stages of tunnels. The "sterilising cooling section" operation mode sterilises the cooling sections. Sterilisation cycle consists of the following steps:

1. Pressure drop
2. Draining heat exchanger
3. Heating up of cooling sections set value of temperature (e.g. 240°C)
4. Sterilisation cooling sections: keeping temperature at recipe set value for recipe set value of time
5. Cooling down without heat exchanger: until temperature reaches < 95°C
6. Cooling down with heat exchanger: until temperature reaches < 25°C

The key requirement of cool zone sterilisation is that the temperature within the zone is maintained at a minimum of 170°C for a period of no less than two hours. This gives a very high degree of assurance that the zone is sterile and suitable for sterile manufacturing operations to occur. In summary, the endotoxin challenge must be sufficient to demonstrate a ≥3 log reduction in endotoxin.

<u>Biological Indicators for Dry Heat</u>

Biological Indicators (BIs) (most commonly Bacillus atrophaeus) are used to demonstrate the efficacy of cool zone sterilisation in depyrogenation tunnels. Using a known indicator population and D-value, the delivered lethality needed to obtain an SAL of at least 10-6 can be determined.
The lethality of a cycle can be calculated using the below equation:
Lethality, $F(h) = \Delta T \times \Sigma L$
$L = 10(t-t_o) / Z$
$Z = 20$ constant
$t_o = 170$ the base temperature (°C)
$t =$ actual temperature (°C)
$\Sigma L =$ cumulative sum of time
$\Delta T =$ time differential (scan time)

<u>Control of Materials</u>

Items intended for sterilisation or depyrogenation should be prepared and maintained under conditions that will ensure that pre–sterilisation or depyrogenation levels of bioburden, particulate and pyrogen contamination are minimised. Items that will come into contact with sterile dosage forms, filling equipment, containers and closures after sterilisation or depyrogenation in a dry heat oven should be packed for sterilisation in an appropriate clean environment. An appropriate standard would be environmental grade C or D under local protection by HEPA-filtered air.

Contamination Considerations

Protection of items against contamination before sterilisation or depyrogenation is not generally an issue when washers and tunnels are integrated.

However, components should in all cases be received in packaging that minimises contamination risk (e.g., from fibres) and handled in such a way as to minimise contamination Items should be clearly identified and controlled to avoid mix-ups between sterile and non-sterile items. Chemical indicators may be attached to containers or placed within loads.

These indicate, through a colour change, that items have been exposed to steriliser conditions but cannot be taken as proof of the adequacy of sterilisation cycles. However, if chemical indicators do not change colour they should be interpreted as confirming sterilisation failure. Items that are not dried immediately after cleaning should be sterilised as soon as possible (no longer than eight hours and preferably within four hours of cleaning) to minimise the risk of microbial proliferation and eventually pyrogen formation between cleaning and sterilisation.

A maximum storage time before re-sterilisation should be specified in case the equipment is not used immediately. Adequate cleaning, drying, and storage of equipment provide for control of bioburden and prevent contribution of endotoxin load.

Start-up Conditions

In sterilising ovens, following any drying phase, the load is typically heated up by closing the dampers to the fresh air supply. Air within the oven is continually recirculated over heating elements and through HEPA filters. In modern sterilisers the cycle is usually under automatic control. If the steriliser requires manual intervention for adjustments (e.g. dampers), then this should be very clearly and precisely defined in the operating SOP and details recorded on the record of each sterilisation cycle. In tunnels, the heating occurs as the components progress into the heating zone.

Control and monitoring should be independent and operate from different temperature sensors. Normally, temperature control and routine monitoring is by fixed position chamber sensors. The relationship to load temperature is established by the validation. If there are movable permanently installed temperature sensors, then these should be placed within the oven chamber and within the most difficult to heat position of the load as determined during validation.

This should be very clearly and precisely defined in the operating SOP and details recorded on the record of each sterilisation cycle. Where only data from fixed position chamber sensors are available, the chamber sensor should be positioned in the same position as used in the validation, generally the most difficult to heat position of the chamber. An appropriate allowance for lag phase should be included in the standard cycle (e.g. to set the steriliser timer).

This approach is used to compensate for load lag times (the time difference between chamber probes and the load cold point reaching sterilising temperature) as established during validation as part of performance qualification. Note that this correction for lag should be part of the standard cycle as defined by validation and incorporated into the operating SOP, included in the automatic or manual control (as applicable) and included as part of the master process record (mpr) or acceptance criteria used to assess the cycle.

Tunnel control and monitoring should be independent, and operate from different temperature sensors.
The control and routine monitoring of tunnels is by fixed position sensors. The relationship between tunnel and load temperature is established by the validation. The tunnel sensors should be positioned in the same position as used in the validation. The acceptance criteria for the cycle are set on the basis of the validation data such that the tunnel load receives the correct heat input. For tunnels, the sterilisation process is continuous and so the temperature record is of a set temperature. Thus, in order to verify heat treatment of components, the belt speed should be confirmed and, if adjustable, recorded either continuously or intermittently (at least at start and end during each day of operation).

In-Process Controls

Process parameters that are essential to sterility assurance should be verified and documented for every load processed. Other less critical process parameters that may be indicative of actual or potential steriliser failure should be verified at a lower frequency. Periodic checks:

- Confirmation of instrument calibration and performance of any applicable calibration checks
- Data to be obtained, documented and verified for each cycle
- Identification of the contents of the load
- Confirmation of compliance with validated loading pattern (ovens only)
- Confirmation of correct sealing of doors

> Confirmation of correct differential pressures
> Continuous record of the temperature, time, belt speed where applicable, throughout each cycle from at least one sensor

Cooling

Oven loads are generally cooled by switching off the heating elements and opening the fresh air dampers, which allows cool HEPA-filtered air to circulate around the load. The rate of cooling should be a compromise between rapidity and the need to avoid product damage. In particular, glass components may be adversely affected by internal stresses caused by rapid and uneven cooling. Note that the cooling phase is established in the validation and fixed as part of the standard cycle.

The checks on cycle records are vital as a failure of sterilisation or depyrogenation cycles may not be readily detectable in the product testing as there are no visible or practicable non-destructive means of testing for sterility.

The assurance of sterility is thus very heavily based upon the validated process conditions being consistently reproduced during routine operation. It is essential that any failures are promptly detected and that there is a clearly defined course of action in the appropriate operating procedure. Any cycle that does not meet any of its acceptance criteria should be thoroughly investigated.

Materials processed through such a cycle cannot be released solely on passing a test for sterility. Any abnormal or unusual occurrences should be formally recorded on the appropriate site documentation and notified to production management and quality management (even if the occurrence is not formally part of acceptance criteria). They should then be assessed for impact on the sterilisation or depyrogenation and on the functionality of the unit. Procedures should be in place to address such situations (e.g. containment measures). There must be a formal, thorough and fully documented investigation of all cycle failures under the site failure investigation procedure. Possible causes of sterilisation or depyrogenation failure(s) include but are not restricted to:

> Components held for insufficient time during sterilisation
> Too low of a depyrogenation temperature in the hot zone

This may happen in the event of the load lag time being longer than expected due to use of unapproved load patterns, over-loaded ovens, inadequate drying etc.

> Ingress of non-sterile air due to inadequate over-pressure, faulty door seals or filter failure

- Dampers failing to operate correctly
- Excessive residual water in containers (from washing stage

Validation

Equipment Installation and Operational Qualification IQ/OQ

Establishing documented evidence that all key aspects of the process equipment installation adhere to the manufacturer's approved specifications and any recommendations of the supplier of the equipment are suitably considered.

The process/equipment operates as intended and all user requirements are adequately fulfilled.

Operational Qualification

The second element of equipment qualification now must be considered; equipment-operational qualification. This is "Establishing by documented evidence that the equipment operates per specifications and over the required ranges and to required tolerances". Equipment is also tested to ensure alarms and controls operate as required and intended. Some typical checks included in an equipment-operational qualification are testing of alarms, control system testing, utility failures and functional and operational testing.

The Validation Lifecycle

The Validation lifecycle refers to the requirement to control and document all validation activities from conception and URS stage to the retirement of equipment or a process. The lifecycle approach ensures compliance throughout the life of the process/equipment while maintaining a validated state throughout the application of change control.

Types of Validation

With most equipment, systems and processes it is best practice to complete all qualification and validation activities in advance of the manufacture of any products for sale, commercial use and use in certain trials. The FDA provides clear definitions on the four types of validation which are explained below.

Prospective Validation
Establishing documented evidence **in advance** of process implementation that a process or system operates as intended. This is the preferred approach and is most common when new products must be validated before commercial manufacturing.

Concurrent Validation
Establishing documented evidence that a process operates as intended, based on information generated during process implementation. Concurrent means that the outputs and performance of the system are monitored at the time of manufacturing which can include commercial lots.

Retrospective Validation
Retrospective validation is used for facilities or processes that have not completed formal validation. Historical data or a retrospective review can provide the evidence that the process or facility is operated as intended. This type of validation is uncommon.

Revalidation
Revalidation involves the re-execution of validation activities in order to maintain a validated state. This can be a result of substantial changes to product attributes, specification or changes to the manufacturing process itself. Other reasons a partial or full revalidation may be required involve instances where product quality issues have increased.

Process Validation

Per FDA 21 Code of Federal Regulations process validation is a regulatory requirement of Good Manufacturing Practices (GMP) for both pharmaceuticals (21 CFR 211) and medical devices (21 CFR 820). In addition to the regulatory drivers, process validation is a requirement in order to obtain certification to international standards issued by many notified bodies. (E.g. ISO 13485 Medical Devices – Quality Management Systems, ASTM E2500-Standard Guide for Specification, Design, and Verification of Pharmaceutical and Biopharmaceutical Manufacturing Systems and Equipment etc.)

What Is Process-Operational Qualification (OQ-P)?

The ability of a process to produce product in accordance with pre-determined specifications under worst case conditions. PQ is only required if no worst-case conditions are evident.

What Is Process-Performance Qualification (PQ)?

The ability of a process to consistently produce product in accordance with predetermined specifications under anticipated conditions (normal/routine conditions). Before considering process validation in further detail, it is important to look at the prerequisites and other supporting activities required. These are examined in the sections below.

It is important to consider test methods early on in the validation life cycle. Before you can begin to consider process validation, test methods should be understood and in place.

A test method is a process or an action used to verify that a product feature meets a predefined specification. Tests methods can be physical or analytical in nature. Test method validation should be completed in advance of process validation to allow the proper assessment of process and product outputs meaning it is often a pre-requisite to process validation.

Examples of test methods include simple visual inspection by microscope, measurement of a dimension with a callipers or measurement of a dimension using an automated optical inspection system. Some test methods will involve MSA (Measurement System Analysis) studies, for example, a measurement of a dimension by an operator using a microscope. In contrast, a test method to determine organic residuals would require an analytical test method validation.

The equipment must be qualified (installation qualification and operational qualification) before the method is validated. Remember – testing completed in contract laboratories or specialist services also require validation! Test methods are critical to the success and integrity of process validation as they assess the outputs. E.g. what are the dimensions, physical attributes or chemical properties of the product and how do they conform to specifications?

The most important point when it comes to validation is that validation is neither exploratory nor investigative. Equally, it is not an engineering study. If you are ready to validate a system or process, all of the groundwork must be completed. This means critical parameters must be defined and documented, with technical rationale on why such parameters are critical etc. This body of work is typically done during a process development study or protocol. Process validation is confirming that a process is capable of consistently manufacturing product under anticipated conditions. Remember, validation should be representative of the commercial process, so any issues in process validation will be repeated in commercial manufacturing.

Consistency, a core principle of process validation, is typically demonstrated by producing three batches/runs for a Process Performance Qualification (PPQ). These batches should be representative of normal production i.e. the size of the batch should be typical of commercial volumes. The PQ study should be executed at nominal conditions, (often termed "anticipated conditions") essentially referring to a controlled environment. Controlled material and controlled parameters (CPPs) are required. Nominal settings should be selected for PQ.

Dominance

The concept of dominance is a term used to describe the "influential" or "dominating" effect on a system or process. Typical examples include the injection moulding process, and packaging process. For example, an injection moulding process can be said to have ***material*** as a dominant factor. Batch-to-batch differences of resin or raw material may cause a change to outputs such as the dimensions of a product or component. If dominant factors cannot be identified or understood a "Designed Experiment" (DoE) technique can be used to properly determine them.

Dominance can be categorised into five sections: (1) setup dominance (2) time dominance (3) worker dominance (4) information dominance and (5) component dominance.

Setup Dominance

Setup Dominance - The Process or equipment relies principally on a procedure or process setup. Process should be stable once "set-up".

Examples include ovens and package sealers. With regard to the oven, the setup would generally be controlled by a recipe or program. This program would be selected by the operator through the Human Machine Interface (HMI). The setup with the correct version of the recipe that contains the desired temperatures, times and pressures is therefore a critical input to the process. With regard to the packaging machine (blister packaging), the correct setup for the tooling and program are critical inputs. If setup dominance is significant, it is best practice to have three separate set-ups/changeovers in the Performance Qualification (PQ).

Time Dominance

The Process or equipment is subject to changes over time (drift over time in temperature, solvent cleanliness, tool wear etc.) The process may need a schedule of process checks and adjustments to ensure process consistency. Examples include CNC Machinery (tool wear) or aqueous based cleaning systems. The tool may only be able to manufacture 1000 parts before defects or quality issues are encountered. If time dominance is significant, three time-points or cycles of expected variation should be made e.g. three points in the cycle (start, middle and end) or three points in a shift (start of shift, middle of shift and end of shift).

Worker Dominance

For worker dominance, the process requires operator experience and skill. Examples include manual or hand finishing. If dominance is significant, ensure there are a minimum of three operators involved in the manufacturing/ activity.

Information Dominance

With information dominance, the process or equipment requires the transmission and/or analysis of information. Examples include LIMS, MRP and ERP systems. A minimum of three information transmissions in the PQ should be completed.

Component Dominance

The process is influenced by the variability of the input materials and/or components. It requires robust inspection and sorting procedures as well as process adjustments. When component dominance is significant, ensure there are a minimum of three component/raw material batches in the PQ sampling plan. If component dominance is significant, this can be mitigated by including the material/component variation in "worst case" testing as part of the Operational Qualification Process (OQ-P)

Process Operational Qualification (OQ-P)

During the Operational Qualification-Process (OQ-P) study, worst-case process conditions are normally employed. This may be worst case temperatures, speeds, feeds etc. The OQ-P should challenge the manufacture/processing of product at the limits of the processing window. If no worst-case conditions exist, then an OQ may not be required and only a performance qualification is required. A family or matrix approach is often used where similar products are to be validated. A particular product size or product configuration may be selected to represent the worst-case product. Therefore, by qualifying the worst case, all other products within that family of products would be considered validated. However, this approach must be clearly documented and technical rationale provided in advance of any qualification activities. This can be addressed in a validation plan or within a protocol.

Process Performance Qualification

The purpose of the PPQ is to demonstrate the capability of the process to consistently manufacture product to pre-determined specifications under normal operating conditions and defined parameters.

Validation is confirmation, so process validation is confirming that a process is capable of consistently manufacturing product under anticipated conditions.
- Lots should be produced consecutively (in sequence)
- Lots must meet the acceptance criteria set out in the protocol
- The lot size should be reflective of the intended lot size and also take into account normal variation
- If a family approach or matrix approach is used, the product selection must be clearly justified and documented
- Execute under anticipated conditions; essentially this refers to a controlled environment. Controlled material, controlled parameters (CPPs)
- Nominal settings should be selected for PPQ

Cleaning Validation

Introduction

Cleaning is the process of removing potential contaminants from process equipment and maintaining the condition of equipment so that it can be safely used for subsequent product manufacture. It is complicated by many different chemicals used to produce medicinal drug products and other chemical agents used in the manufacturing process or in the cleaning process.

Key regulatory and international publications are included below:

- FDA – Food and Drug Administration – Guide to Inspections of Validation of Cleaning Processes

- EU GMP – European Commission – EudraLex Volume 4: EU Guidelines to Good Manufacturing Practice, Medicinal Products for Human and Veterinary Use, and Annex 15 (section 10 "Cleaning Validation")

- ICH Q7 – International Council on Harmonisation - Good Manufacturing Practice

- Guide for Active Pharmaceutical Ingredients (section 12.7 "Cleaning Validation")

- ICH Q9 – International Council on Harmonisation – Quality Risk Management

- PIC/S PI 006-3 – Pharmaceutical Inspection Co-Operation Scheme – Recommendations on Validation Master Plan, Installation and Operational Qualification, Non-Sterile Process Validation, Cleaning Validation (section 7 "Cleaning Validation")

- WHO TRS 937 – World Health Organisation - Specifications for Pharmaceutical Preparations; Annex 4: Supplementary guidelines on Good Manufacturing Practices: Validation; Appendix 3: Cleaning Validation

FDA – Food and Drug Administration - Guide to Inspections of Validation of Cleaning Processes

Cleaning validation programmes are important requirements for both bulk pharmaceutical processing and biotechnology. As with validation of other processes, there may be more than one way to validate a cleaning process. Once the manufacturer can establish inspection consistency and repeatable outcomes that ensure predetermined acceptable criteria are met, a cleaning procedure can be deemed effective. This is a driven process which should support claims of consistent outcomes.

It isn't solely the FDA that has an expectation that cleaning procedures (processes) be validated; PIC/s, ICH, EudraLex and WHO guidance and requirements also specify the need to validate cleaning procedures.

Historically, the FDA was mostly concerned about the contamination of non-penicillin drug products with penicillin or the cross-contamination of drug products with potent hormones or steroids. One event which increased FDA awareness of the potential for cross-contamination due to inadequate procedures was the 1988 recall of a finished drug product, Cholestyramine resin USP.

In this instance, the bulk pharmaceutical used to produce the product had become contaminated with low levels of both intermediates and degradants.

The cross-contamination in this case was attributed to the reuse of recovered solvents. The recovered solvents had been contaminated because of a lack of control over the reuse of solvent drums. Drums that had been used to store recovered solvents from a pesticide production process were later used to store recovered solvents used for the resin manufacturing process. Some shipments of this pesticide-contaminated bulk pharmaceutical were supplied to a second facility at a different location for finishing. This resulted in the contamination of the bags used in that facility's fluid bed dryers with pesticide contamination.

The main focus of an auditor in respect of cleaning validation is to evaluate the evidence that aims to demonstrate the effectiveness of the approach and processes used to clean equipment.

The following questions are relevant when evaluating the cleaning process:
- At what point does a piece of equipment/system become clean?
    - This knowledge should be captured in cycle development and development of the cleaning process. Studies may indicate that a vessel or piece of equipment requires three rinses with hot water-for-injection at which point it meets acceptance criteria. However, an additional number or rinses may be included to provide a level of confidence in the cleaning process.
- Does it have to be scrubbed by hand?
    - Depending on the drug substances and excipients or other chemicals, residues may tend to physically "stick" to surfaces or behave as tarry or gummy which may require mechanical force to remove them, or a solvent rinse may be sufficient for removal.

When the cleaning process is used only between batches of the same product, a company may only meet the criteria of "visibly clean" for the equipment. This can often be referred to as a batch-to-batch clean. Such between-batch cleaning processes do not require validation. Change-over from one product to a different product of different materials requires a more comprehensive clean, potentially requiring multiple cleans or rinses.

EU GMP – European Commission – EudraLex Volume 4: EU Guidelines to Good Manufacturing Practice, Medicinal Products for Human and Veterinary Use, and Annex 15 (section 10 "Cleaning Validation")
Section 10 of Annex 15 provides a number of bullet points with regard to cleaning validation:

"Cleaning validation should be performed in order to confirm the effectiveness of any cleaning procedure for all product contact equipment. Simulating agents may be used with appropriate scientific justification. Where similar types of equipment are grouped together, a justification of the specific equipment selected for cleaning validation is expected.
A visual check for cleanliness is an important part of the acceptance criteria for cleaning validation. It is not generally acceptable for this criterion alone to be used. Repeated cleaning and retesting until acceptable residue results are obtained is not considered an acceptable approach.
It is recognised that a cleaning validation programme may take some time to complete and validation with verification after each batch may be required for some products, e.g. investigational medicinal products. There should be sufficient data from the verification to support a conclusion that the equipment is clean and available for further use.
Validation should consider the level of automation in the cleaning process. Where an automatic process is used, the specified normal operating range of the utilities and equipment should be validated.
For all cleaning processes an assessment should be performed to determine the variable factors which influence cleaning effectiveness and performance, e.g. operators, the level of detail in procedures such as rinsing times etc. If variable factors have been identified, the worst case situations should be used as the basis for cleaning validation studies.
Limits for the carryover of product residues should be based on a toxicological evaluation. The justification for the selected limits should be documented in a risk assessment which includes all the supporting references. Limits should be established for the removal of any cleaning agents used. Acceptance criteria should consider the potential cumulative effect of multiple items of equipment in the process equipment train.
The risk presented by microbial and endotoxin contamination should be considered during the development of cleaning validation protocols.
The influence of the time between manufacture and cleaning and the time between cleaning and use should be taken into account to define dirty and clean hold times for the cleaning process.
Where campaign manufacture is carried out, the impact on the ease of cleaning at the end of the campaign should be considered and the maximum length of a campaign (in time and/or number of batches) should be the basis for cleaning validation exercises.
Where a worst case product approach is used as a cleaning validation model, a scientific rationale should be provided for the selection of the worst case product and the impact of new products to the site assessed. Criteria for determining the worst case may include solubility, cleanability, toxicity and potency.
Cleaning validation protocols should specify or reference the locations to be sampled, the rationale for the selection of these locations and define the acceptance criteria.

Sampling should be carried out by swabbing and/or rinsing or by other means depending on the production equipment. The sampling materials and method should not influence the result. Recovery should be shown to be possible from all product contact materials sampled in the equipment with all the sampling methods used.

The cleaning procedure should be performed an appropriate number of times based on a risk assessment and meet the acceptance criteria in order to prove that the cleaning method is validated.

Where a cleaning process is ineffective or is not appropriate for some equipment, dedicated equipment or other appropriate measures should be used for each product as indicated in chapters 3 and 5 of EudraLex, Volume 4, Part I.

Where manual cleaning of equipment is performed, it is especially important that the effectiveness of the manual process should be confirmed at a justified frequency."

Ref: EU GMP V4, Annex 15, 2017

ICH Q7 – International Council on Harmonisation - Good Manufacturing Practice:

"Cleaning procedures should normally be validated. In general, cleaning validation should be directed to situations or process steps where contamination or carryover of materials poses the greatest risk to API quality. For example, in early production it may be unnecessary to validate equipment cleaning procedures where residues are removed by subsequent purification steps. (12.70)

Validation of cleaning procedures should reflect actual equipment usage patterns. If various APIs or intermediates are manufactured in the same equipment and the equipment is cleaned by the same process, a representative intermediate or API can be selected for cleaning validation. This selection should be based on the solubility and difficulty of cleaning and the calculation of residue limits based on potency, toxicity, and stability. (12.71)

The cleaning validation protocol should describe the equipment to be cleaned, procedures, materials, acceptable cleaning levels, parameters to be monitored and controlled, and analytical methods. The protocol should also indicate the type of samples to be obtained and how they are collected and labelled. (12.72)

Sampling should include swabbing, rinsing, or alternative methods (e.g., direct extraction), as appropriate, to detect both insoluble and soluble residues. The sampling methods used should be capable of quantitatively measuring levels of residues remaining on the equipment surfaces after cleaning. Swab sampling may be impractical when product contact surfaces are not easily accessible due to equipment design and/or process limitations (e.g., inner surfaces of hoses, transfer pipes, reactor tanks with small ports or handling toxic materials, and small intricate equipment such as micronisers and microfluidisers). (12.73)

Validated analytical methods having sensitivity to detect residues or contaminants should be used. The detection limit for each analytical method should be sufficiently sensitive to detect the established acceptable level of the residue or contaminant. The method's attainable recovery level should be established. Residue limits should be practical, achievable, verifiable, and based on the most deleterious residue. Limits can be established based on the minimum known pharmacological, toxicological, or physiological activity of the API or its most deleterious component. (12.74)

Equipment cleaning/sanitation studies should address microbiological and endotoxin contamination for those processes where there is a need to reduce total microbiological count or endotoxins in the API, or other processes where such contamination could be of concern (e.g., non-sterile APIs used to manufacture sterile products). (12.75)

Cleaning procedures should be monitored at appropriate intervals after validation to ensure that these procedures are effective when used during routine production. Equipment cleanliness can be monitored by analytical testing and visual examination, where feasible. Visual inspection can allow detection of gross contamination concentrated in small areas that could otherwise go undetected by sampling and/or analysis. (12.76)"

PIC/S PI 006-3 – Pharmaceutical Inspection Co-Operation Scheme – Recommendations on Validation Master Plan, Installation and Operational Qualification, Non-Sterile Process Validation, Cleaning Validation (section 7 "Cleaning Validation")

PIC/s provides several pages of recommendations on cleaning validation. It clearly outlines the principles and purpose of conducting cleaning validation:

"Pharmaceutical products and active pharmaceutical ingredients (APIs) can be contaminated by other pharmaceutical products or APIs, by cleaning agents, by micro-organisms or by other material (e.g. air-borne particles, dust, lubricants, raw materials, intermediates, auxiliaries). In many cases, the same equipment may be used for processing different products. To avoid contamination of the following pharmaceutical product, adequate cleaning procedures are essential.

Cleaning procedures must strictly follow carefully established and validated methods of execution. This applies equally to the manufacture of pharmaceutical products and active pharmaceutical ingredients (APIs). In any case, manufacturing processes have to be designed and carried out in a way that contamination is reduced to an acceptable level.

Cleaning validation is documented evidence that an approved cleaning procedure will provide equipment which is suitable for processing of pharmaceutical products or active pharmaceutical ingredients (APIs).

The objective of cleaning validation is confirmation of a reliable cleaning procedure so that the analytical monitoring may be omitted or reduced to a minimum in the routine phase."

Similarly, cleaning validation can be divided into the same three stages; cleaning process design, cleaning process validation and continued process monitoring of cleaning processes. Prior to the commercial manufacture and distribution of drug products and medicines to consumers, a manufacturer must gain a high degree of assurance in the performance of the manufacturing process such that it will consistently produce APIs and drug products meeting those attributes relating to identity, strength, quality, purity and potency. A high degree of assurance and demonstrated consistency of a process must be evidence-based.

PIC/S Guidance on Limits

The Pharmaceutical Inspection Convention and Pharmaceutical Inspection Co-operation Scheme (jointly referred to as PIC/S) are two international instruments between countries and pharmaceutical inspection authorities which provide together an active and constructive co-operation in the field of GMP.[1]

The most important point to remember when it comes to limits is that residues meet predefined criteria, the most stringent criteria as listed below:

(a) No more than 0.1% of the normal therapeutic dose of any product should appear in the maximum daily dose of the following (next) product,
(b) No more than 10 (parts per million, ppm) of any product will appear in another product, (this value is not always the default)
(c) No quantity of residue should be visible on the equipment after cleaning procedures are completed. Spiking studies should determine the concentration at which most active ingredients are visible. [2]

The method of determining residue limits of active ingredients is based on an approach developed by Fourman and Mullen (1993) and is referenced in PIC/s guidance amongst other publications.

Test Methods

It is important to consider test methods and test method validation early on in the validation life cycle. A test method is a process or an action used to verify that a product feature or particular requirement meets a predefined specification. Test methods can be physical or analytical in nature. Test method validation should be completed in advance of cleaning as the test method is used to verify the outputs of such cleaning validations.

ICH, Q7, Validation of Analytical Methods

"Analytical methods should be validated unless the method employed is included in the relevant pharmacopeia or other recognised standard reference. The suitability of all testing methods used should nonetheless be verified under actual conditions of use and documented. (12.80)
Methods should be validated to include consideration of characteristics included within the ICH guidance on validation of analytical methods. The degree of analytical validation performed should reflect the purpose of the analysis and the stage of the API production process. (12.81)
Appropriate qualification of analytical equipment should be considered before initiating validation of analytical methods. (12.82)

---

[1] http://www.picscheme.org/

[2] http://www.picscheme.org/publication.php?id=4

Complete records should be maintained of any modification of a validated analytical method. Such records should include the reason for the modification and appropriate data to verify that the modification produces results that are as accurate and reliable as the established method. (12.83)
(Reference: Q7 Good Manufacturing Practice Guidance for Active Pharmaceutical Ingredients Guidance for Industry September 2016.)

Materials of Construction

When it comes to materials of construction, the same selection criteria can be applied to precision cleaning systems and CIP equipment trains. Above all, materials and their surfaces should be non-reactive, non-corrosive and non-porous. Stainless steel of a high grade is often the preferred material of construction. Examples of grades used include 304, 316 and 316L. For surfaces that are product contacting, material certificates are required to provide evidence that the materials and their constituents are of the correct make-up and suitable grade.

Stainless steels (SS) are crystallised solutions of at least 11% of corrosion reducing elements like chromium and nickel in iron. Generally, they are iron based with 12 to 30% chromium, 0 to 22% nickel and minor or no amounts of carbon, columbium, copper, manganese, molybdenum, nitrogen, phosphorus, selenium, silicon, sulfur, tantalum, and titanium.

Casting grades generally are designed with more sulphur to facilitate welding and have more ferrite (a less corrosion resistant phase) to prevent the formation of micro-cracks on cooling.

Preferred stainless steels for use in the life sciences are manufactured by VIM – vacuum-induction-melt followed by a secondary VAR – vacuum-arc-re-melt process with sulphur add-back and dispersion in order to minimise inclusions (stringers) and control the amount of sulphur used.

The surface of stainless steel can also be contaminated with the electrolyte solution used in electro-polishing if it is used an excessive amount of times or if rinsing steps are not adequate.

This solution builds up iron and other contaminants that can be transferred to the part being electro-polished if the conditions and chemistry are not carefully controlled.

To prevent these problems from occurring, all electrolyte solutions must be removed from the surface by using a chemically pure water rinse until the conductivity of the rinse from the stainless steel is equal to the conductivity of the water being supplied for rinsing.

Pressure Testing

Piping or system integration can be required for:
- Precision cleaning systems where the utilities need to be "tied" in to the system

- Installation of a pharmaceutical process within a facility e.g. a skid[3] plug in.

---

3 see definition in introductory pages

After installation, (and before passivation if required) piping systems are pressure tested by filling the system with clean air to 150% of the design pressure or 150psi, whichever is the greatest value. The pressure is then monitored over a 4-hour period to see if there is any drop in pressure.

Passivation

Passivation can be described as the active chemical process used to obtain a uniform chromium oxide layer on stainless steel (SS) surfaces. The chromium oxide layer or film forms a protective coating that gives corrosion resistant properties.

The protective layer naturally forms from the reaction of oxygen in air with the chromium on the metal surface but this naturally forming layer can be non-uniform or patchy due to impurities and surface chemistry defects.

When stainless steel is worked such as in welding, machining, mechanical polishing etc., its uniformity of the naturally forming protective layer can be damaged and oxides of other compounds forming the stainless steel composition can occur. Corrosion can begin at these non-uniform sites and, because stainless steel contains over 60% iron, the corrosion can proliferate from the surface through the body of the metal if no opportunity for protective layer reformation is given.

There are three passivation processes that are used to enhance the corrosion resistance of stainless steel:

- Treatment with oxidising acids
- Treatment with chelants
- Treatment by electro-polishing

Layer formation is a dynamic chemical process where the chromium atoms are combined with oxygen (and hydroxyl ions in aqueous environments) to form a complex surface layer that prevents attack on corrosion-prone atoms such as iron. Nickel and molybdenum may play a role in formation of the passive film but the mechanism has not been proven.

Sampling

There are two main methods of sampling that are considered to be acceptable; direct surface sampling and indirect sampling (use of rinse solutions). A combination of the two methods is generally the most desirable, particularly in circumstances where accessibility of equipment parts can mitigate against direct surface sampling.

Direct Surface Sampling

The suitability of the material to be used for sampling and of the sampling medium should be determined. The ability to recover samples accurately may be affected by the choice of sampling material. It is important to ensure that the sampling medium and solvent are satisfactory and can be readily used. Ref: PIC/S PI 006-3.

Rinse Samples

Rinse samples allow sampling of a large surface area. In addition, inaccessible areas of equipment that cannot be routinely disassembled can be evaluated. However, consideration should be given to the solubility of the contaminant.

A direct measurement of the product residue or contaminant in the relevant solvent should be made when rinse samples are used to validate the cleaning process. Ref: PIC/S PI 006-3.

In rinse sampling a fluid (solvent) is used to rinse and make contact with all surfaces of the item. The sample is then tested quantitatively to remove the target residue.

Sampling aims to detect any residue drug content or solvents or other soiling left behind after the cleaning process. The visual inspection is also important in identifying any larger contamination of debris. Microbial sampling is also done to ensure no microorganisms are present in equipment or in areas of production.

Microbial Sampling

Microbial sampling of utilities such as water-for-injection, purified water, process air etc. is required to ensure no bacteria, moulds, fungi or yeasts are present which risk patient safety. Colonies can often be determined by visual inspection based on the attributes and appearance test plates/samples. If visual identification is not possible, the colony should be sent for gram stain analysis.

Visual checks involve assessing plates for:
- colour
- shape
- elevation
- size
- texture
- surface
- edge appearance

The key utilities involved for cleaning include utilities such as water, compressed gases (air, nitrogen etc.) and the heating and cooling of process equipment. Water quality can impact the effectiveness of pre-rinsing, washing, and final rinsing.

Therefore, both the water temperature and quality need to be tightly controlled and monitored. Gases are typically used in order to blowdown or blowout remaining fluids or they are used as a drying step.

The term "clean utilities" in the life science industry refers to utilities that have to fulfil regulatory requirements. The most common utility is water, which can be supplied in different pharmaceutical grades of purity. Purified water (PW or PUW), Highly

Purified Water (HPW) and Water-for-Injection (WFI) are the most common. Water quality specifications can be found in the pharmacopeias, e.g. the US Pharmacopeia. Other clean utilities can also include clean compressed air, clean gasses (e.g. nitrogen, argon and oxygen), and clean steam.

www.ingramcontent.com/pod-product-compliance
Lightning Source LLC
Chambersburg PA
CBHW070421220526
45466CB00004B/1491